Detail *Res Materia* 2017_38 by Sanne Karssenberg

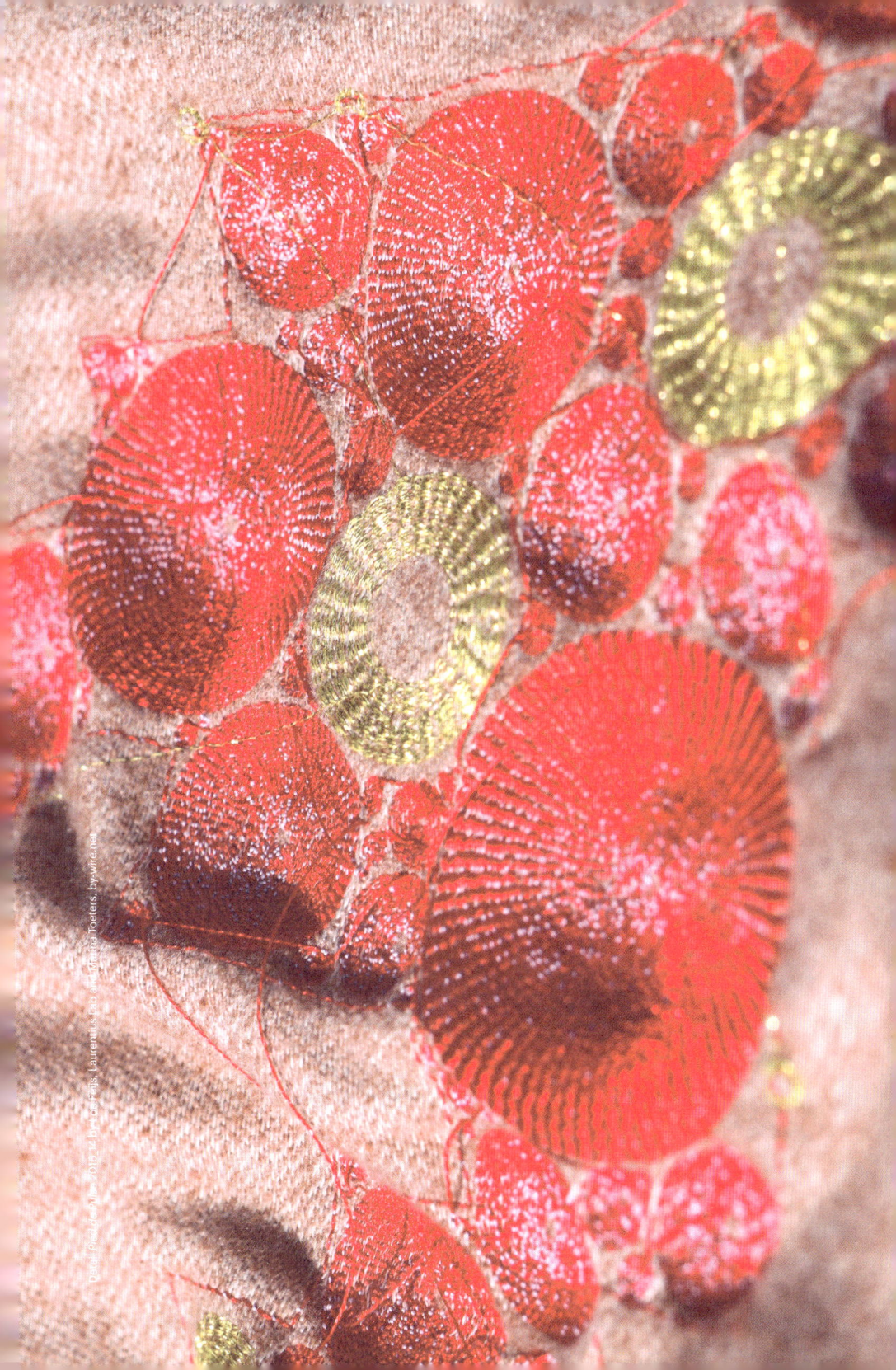

Unfolding Fashion Tech:

Produced and edited
by Marina Toeters

Pioneers of Bright Futures

2000_2020

Pioneers
(in order of appearance)

2000-09_01
Philips

2007_02
by-wire.net, Marina Toeters

2009_03
Melissa Coleman

2009_04
by-wire.net, Marina Toeters JSSSJS, Jesse Asjes

2009-14_05
Anja Hertenberger, Barbara Pais and Danielle Roberts

2009-16_06
Gail Kenning

2010_07
ESA and by-wire.net

2010_08
University of the Arts Utrecht, Tim Walther and by-wire.net, Marina Toeters

2010-14_09
Philips Research

2012_10
Laurentius Lab, Philips Research and by-wire.net

2012_11
m.nster. and Studio Roosegaarde for Lacoste

2012_12
Contre Choc and by-wire.net

2012-15_13
Meg Grant, Ralf Jacobs, Marina Toeters and Aniela Hoitink

2013-17_14
Laurentius Lab, Loe Feijs and by-wire.net, Marina Toeters

2013-15_15
TU/e Wearable Senses Lab and many others

2014_16
Pauline van Dongen

2014_17
MVO consortium including by-wire.net

2014_18
Tamara Hoogeweegen

2014-15_19
Anke Jongejan

2015_20
Holst Centre and by-wire.net

2015_21
Bruna Goveia Da Rocha

2015_22
Jasna Rok

2015_23
Eef Lubbers and Malou Beemer

2015_24
Aniela Hoitink, NEFFA

2015_25
by-wire.net, Marina Toeters and Martijn ten Bhömer

2015_26
Pauline van Dongen

2015_27
Qi Wang, Eindhoven University of Technology

2015_28
Karin Vlug and Laura Duncker

2015_29
Leonie Tenthof van Noorden and Eunbi Kim

2015-18_30
Mohamad Zairi Baharom

2016_31
Anouk Wipprecht

2016-19_32
Bambi Medical, MedTech and Sibrecht Bouwstra

2016_33
Ilja Visser and by-wire.net

2016_34
Spellbound

2016_35
Saxion

2017_36
Pauline van Dongen

2017_37
Brigitte Kock, Bart Pruijmboom and Niek van Sleeuwen

2017_38
Sanne Karssenberg

2017_39
Lithe Lab, Daisy van Loenhout

2017_40
Philips

2018_41
Beam Contre Choc

2018_42
Fabienne van der Weiden and Jessica Joosse

2018_43
Laura Luchtman and Ilfa Siebenhaar

2018_44
Bianca Gorini

2018_45
Holst Centre, StudioBonvie and by-wire.net

2018_46
Puck Martens, Merle Kroezen and Suzanne Mulder

2018_47
Hellen van Rees

2018_48
Kristi Kuusk

2018_49
Karin Vlug and Bas Froon

2019_50
Angella Mackey

contents

What Happens in the Field of Fashion Innovation?
And Why This Book?

foreword

By Marina Toeters

Fashion has a reputation as a site of innovation, and for embracing the novel and the new. Each change of season marks the launch of new collections while discussion turns to the 'latest fashion trend', However, despite technological advances in new materials, design processes, artificial intelligence, 3D-printing and design software, changes in the fashion industry and everyday apparel have not been as significant as we might expect. Mid-twentieth century science fiction programs foresaw shape-shifting fabrics, technologies and designs for the future that have never come to pass. Synthetic textiles promised to change the world. But when we look at the clothes worn on the street it's hard to believe so little has changed since the 1950s.[1]

Fashion is at an impasse, but the existing fashion industry faces growing criticism. The world needs an alternative. "Fashion presents itself as a glamorous industry, but that is only the tip of the iceberg. Underneath, there is sweat, exploitation, pollution, secrecy and a massive obsolescence industry, namely a constant drive to make existing forms look outdated and to create new forms or novelties."[2] A growing number of designers are exploring alternatives. And, if fashion would embrace just some of these alternative approaches, the industry could change completely within just one or two seasons. This book aims to show what the future of fashion could look like. It explores alternatives to the plentitude of discarded garments buried in landmasses and offers a hopeful vista for the future of textiles and technology. We sincerely believe that this is the true power of fashion.

The fashion innovation arena is not a single homogeneous field. It is a junction point of people, designers and companies from different disciplines, pioneers driven to innovate clothing for the better. The Dutch city of Eindhoven is a design and technology hub and plays a leading role in design education and innovation. Each year the Dutch Design Week (DDW) exposition showcases developments in design. In DDW 2018, the exhibition

foreword_What Happens in the Field of Fashion Innovation?
And Why This Book?

15

'Fashion? Future design for the present' explored a wide range of possibilities for a more sustainable industry highlighting a range of innovative fashion projects from both Dutch and international designers and researchers. Some projects were in experimental stages, while others were market-ready and available for home use. The exhibition was a great success and visitors were keen to get more information about the project. This book aims to expand on this exhibition and make a statement about Dutch designers and researchers, to show what we have done, where we are going, and to introduce the people and the ideas that can take us forward.

I am Marina Toeters, the driving force behind this publication. Through my business I stimulate collaborations between the fashion industry and technicians to generate change in the fashion system and to develop supportive garments for everyday use. Since 2006, I have operated on the cutting edge of fashion technology and fashion design with my studio by-wire.net. by-wire.net has many different objectives. It is dedicated to the design and prototyping of innovative textile products and garments in cooperation with industry giants such as Philips Research and Holst Centre. by-wire.net engages in complex research and development projects and invests in the future of fashion by speaking at events and through the education of young designers. As a teacher, coach and researcher, I work for leading organisations in design, including the fashion department at University of the Arts Utrecht (HKU) and the Industrial Design faculty at the Eindhoven University of Technology (TU/e). I passionately believe that through by-wire.net, sharing knowledge, educating and developing cutting edge prototypes, we can conceive and actualize a better future for the fashion industry. The future is close. Fashion is powerful—not only is it one of the largest industries in the world—people spend their entire lives in fashion. Let's not waste the power of fashion, but let's innovate for a better world.

This book is a fifteen-year journey through the field of fashion and technology. Personal benchmarks are presented alongside pioneering projects of leading fashion technology designers and show innovative concepts coming to fruition. The *Triangled Coat* 2007_02 produced as part of my Master of Arts study at the University of the Arts

2007_02

2015_25 NazcAlpaca, by-wire.net, Marina Toeters and Martijn ten Bhömer

2018_45 Closed Loop Smart Athleisure Fashion, Holst Centre, StudioBonvie and by-wire.net

2007_02 Triangle Coat, by-wire.net, Marina Toeters

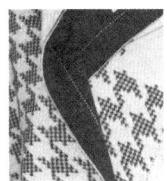

2015_25

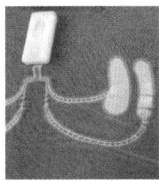

2018_45

Utrecht, fashion design, challenged me to innovate by developing a method for 3D measurement and design. Other projects included the futurist *Human & Kind* project for the European Space Agency [2010_07], the development of garments using Solar Fiber [2012-15_13], the tactile yet utilitarian garments of NazcAlpaca [2015_25], and more recently the Closed Loop Smart Athleisure Fashion [2018_45].

We brought together the theorists, scientists and designers behind the Dutch wearable technologies field to discuss their different perspectives on the challenges faced by those inside and outside the field. The combination of their perspectives provides a bird's eye view of the developments that have already been realized, with the aim to anchor current knowledge, contextualise it and to provide room for reflections on these achievements, allowing us to look to the future with optimism.

In this book we will introduce the reader to the field of wearable technologies and a range of issues that impact fashion now. Chapter one begins with a foundational discussion of the three pillars of fashion and technology design. Professor Daniëlle Bruggeman, Professor Jan Mahy and Rens Tap, three experts from very different disciplines within the fashion field, discuss fashion's cultural context and its relationship with the human body, the technical development of textiles in the Netherlands, and offer insight into the economy of fashion business, current trends, and production modes and locations. In chapter two Professor Ben Wubs continues the analysis of fashion as a business and presents a critical and constructive view on the business of fashion from a historical perspective. He offers a contextual overview of how fashion and technology developed over the last centuries. Chapter three sets up a dialogue between technical researcher Professor Loe Feijs, and high-tech industry expert Koen van Os. They explain how technological research transitioned from the development of e-textile prototypes to wearable products that take their place in society.

In chapter four, cultural researcher Dr. Lianne Toussaint focuses on the socio-cultural implications of the integration of fashion and technology. In chapter five, Dr. Gail Kenning explores how technology and textiles together offers opportunities to promote wellbeing. Professor

foreword_What Happens in the Field of Fashion Innovation?
And Why This Book?

17

Stephan Wensveen continues the discussion in chapter six by discussing the role of education in interaction design, and interdisciplinary research with industry involvement, through the Wearable Senses Lab at Eindhoven University of Technology. Anke Jongejan, in chapter seven, introduces the concept of design fiction and speculative design to show how these approaches provide solutions to the key barriers to innovation in the fashion industry. The separation between technological functionality and fashion aesthetics will be discussed in chapter eight by Pauline van Dongen and Dr Oscar Tomico. They show what it takes to integrate technologies into clothing based on their experience and knowledge of the CRISP Smart Textile Services and Crafting Wearables project. Finally, chapter nine will offer a brief overview before asking 'where to from here?'

Marina Toeters, *Initiator, producer and editor of this publication. Owner of by-wire.net. Designer, educator and researcher in fashion technology.* Marina Toeters operates on the cutting edge of fashion technology and fashion design. Through her business by-wire.net she stimulates collaboration between the fashion industry and technicians for a relevant fashion system and support-ive garments for everyday use. The by-wire.net studo is dedicated to designing and prototyping innovative textile products and garments and advises, amongst others, Philips Research and Holst Centre on product development. As a teacher, coach and researcher, she works for multiple institutes like the fashion department at University of the Arts Utrecht (HKU) and the Industrial Design faculty at the Eindhoven University of Technology (TU/e) .

1
Toeters, M. (2016, September). E-fashion fusionist aiming for supportive and caring garments. *In Proceedings of the 2016 ACM International Joint Conference on Pervasive and Ubiquitous Computing: Adjunct* (pp. 922-926). ACM.

2
Wubs, B. (2019, March 6). Capitalism's Favourite Child. Towards an International Business History of Fashion. *Inaugural Address Prof. Ben Wubs.* https://www.eur.nl/en/eshcc/news/inaugural-address-prof-ben-wubs

2013_15 Vibe-ing. TU/e Wearable Senses Lab and many others

chapter i

I   The First Pillar: Dancing with Fashion
     By Daniëlle Bruggeman [1]

It is time to re-engage with things that matter. Fashion
matters. There are many crises and challenges in society
that fashion could take responsibility for, yet fashion
generally focuses on itself. Fashion has a grandiose Ego
formed by the spectacle of the runway, glamour, star
designers, constructed desire and seduction, money, an
abundance of visual imagery and an excess of consumer
products. The ways in which the current fashion system
operates often denies the lived experiences of those who
wear and make clothes.

Rather than acting from a point of resistance to the
current fashion system, it is important to start with an
affirmative approach. By re-engaging—*dancing*—with
fashion in a different way. Fashion is all about the intimate
relationship between the body and materiality. It is a
continuous process of fabrics, colours, textures acting
one on another. While the fashion system is highly material-
istic, obsessed with consuming material products, the
material physicality of 'things' urgently needs more care.
The current fashion system reduces the physical matter
of fashion to something worthless that is exploited for
commercial aims. In re-engaging with fashion, it is important to draw more attention to the intimate and sensorial
experience of fashion, and to develop an embodied
relationship with fashion.

2013_15

The relationship between the physical body and material
objects of fashion is necessarily sensorial [2013_15]. The human
body is sensorial matter, and the condition of embodiment
is 'the way in which we literally and figurally make sense
of, and to, both ourselves and others'.[2] Our embodied
experience of fashion and of wearing clothes necessarily
engages all the senses. Wearing cloth on the body affects
one's physical, sensorial and embodied experiences.
It is about being moved, touching, being touched. This
affective relationship between human being and material
object is about the personal connection and emotional

bond with fashion, which is experienced sensorially [2007_02]. Acknowledging this emotional dimension is an essential part of moving towards a more sustainable relationship with clothes and a more engaged future for fashion. In the discussion on sustainability and circularity, the emotional value and affective relationship between object, maker and particularly the wearer, deserves much more attention [2015_24].

2015_24

Moreover, the senses are crucial in order for us to imagine the future. To imagine the future as a potential reality, or critically reflect on the future implications of our current actions, it is essential to trigger all the senses.[3] This will help to develop a more sustainable relationship between human beings and the material objects that surround our bodies and living spaces and that mediate our human experience [2017_38].

II   The Second Pillar: High Performance Textiles
        By Jan Mahy

Two high-performance 'made in Holland' fibre innovations in the last decade of the twentieth century have put the Netherlands on the map, and made it a huge player in the area of high-end technical textiles. These fibres are AkzoNobel's aramid fiber Twaron®, now a part of Teijin, and DSM's ultra high molecular weight polyethylene (UHMWPE) fibre Dyneema®. These fibres have been applied in aircrafts, electronics, anti-ballistic jackets and flame-resistant personal protection apparel [2007_02]. Because of this innovation, personal protection apparel for the military (TenCate) [2007_02] and other technical textile applications could stay ahead of the competition.

2007_02

The textile industry has reached a new turning point in the last two decades. The mass market production of natural fiber fabrics in low-wage countries has fast become unsustainable, and workers are working in deplorable conditions. Simultaneously, increased awareness in relation to health and well-being [2010-14_09], the ageing population [2015_27] and the scarcity of natural resources such as water and energy [2007_02], have prompted the development of circular processes. It is predicted that, in the coming decades, an increasing gap will emerge between the demand for new

2016_35 SaXcell, Saxion

2015_24 MycoTEX, NEFFA, Aniela Hoitink

2014_17 Sustainable and Supportive Garments for Nurses, MVO consortium including by-wire.net

2007_02 Fire Fighter Suit, by-wire.net

2007_02 Tecatud, by-wire.net

2016_35

2014_17

garments and the ability to supply virgin raw materials. This drives the need to improve the end-of-life collection of textiles and the separation and re-use of fiber material into fabrics, such as for instance the Econyl® process [2018_45] and the SaXcell® chemical recycling of cotton [2016_35]. To support a circular approach to material use, it is vital to look at innovative product development to ensure products are structured and designed for the circular economy. This can be done by, for example, focusing on zero waste and working with recycled waste materials, such as the Textile Reflexes products [2018_47].

At the interface of the new textile technology development, the response to the garment industry's need for innovative e-textiles and made-to-order customized fabrics, and the regulatory need for circular manufacturing processes, innovation is manifold. Examples are smart materials and e-textiles in healthcare, workwear [2014_17] and sports clothing [2015_26]. Fostered by EU-funded initiatives, consortia of industrial partners across the value chain and academic knowledge centres collaborate to make promising ideas and concepts ready for market entry. The role of Universities of Applied Sciences in the Netherlands, such as Saxion, fulfills not only a regional function to attract talented students, but also helps to bridge the 'valley of death' between proving a concept within a lab environment and the development of an industrially attractive business case.

III   The Third Pillar: Business Strategy for Flow Fashion
      By Rens Tap

During the 1980s a significant amount of fashion was still produced within the Netherlands. However, globalisation and the later development of the Internet had major impact. Fashion producers focused on branding, marketing and retailing. Successful retailers and brands became global players through the verticalization of production, logistics and omni-channel retail strategies. High volumes of fashionable clothes were produced at low costs, a combination that previously had seemed impossible. Small independent brands and retailers were also impacted. They needed to speed up production, adjust their strategies towards e-commerce, and apply digitalisation in long, cumbersome and slow supply chains. Following the 2008 recession, sales of fashion dropped until 2012. Many brands proven to be economically vulnerable, without capital to invest in innovation. The business strategy was risk avoidance in order to survive.

Ten years after the recession, the turnover of the Dutch fashion industry is slowly recovering. But there has been little change in product innovation. E-commerce is booming, but high shop rents and personnel costs remain a heavy financial burden. Returned goods from e-commerce and a shorter life-cycle of textiles and clothes have resulted in unsustainable logistic flows and piles of waste. The call for high efficiency through large volumes caused margins to creep and put many brands and producers out of business. The cost of importing clothing has risen, but the average price of apparel has remained almost the same despite inflation, resulting in an exhaustive race to the bottom. In addition, consumers are increasingly aware of the origins of their clothes. Calls for sustainability and better working conditions are impacting the mainstream. Retailers are at a loss on how to create a transparent supply chain when production is scattered and far away.

2018_46

Fashion brands need to innovate and add new values to their products and services. NL NextFashion and Textiles is a network of the branch fashion and textile organisation (Modint) and four Universities of Applied Sciences in the Netherlands, supported by the Dutch government. The network supports fashion brands in addressing design, digitization [2018_46] and sustainability. While poor recycling processes remain a problem the Waste-Conscious scarf [2012_12] explored the future of circularity, a process that is already applied in the workwear sector [2014_17]. Today, fashion production is once again taking place in and around Europe and supply chains are shortening, due to the rising wages in China. New production methods are being explored that make it possible to produce made-to-measure garments at any time and in any place [2015_25], [2015_29], [2017_37], [2018_42]. New materials are being developed of which only a small selection can be presented in this book [2016_35], [2018_43]. Fast fashion? Slow fashion? In my opinion we have to move towards flow fashion: A continuous flow of sustainable material, and value creation for the wearers!

2015_29

2018_46 Studio PMS

2015_29 This Fits Me,
Leonie Tenthof van Noorden
and Eunbi Kim

Daniëlle Bruggeman, *Professor of Fashion, ArtEZ University of the Arts*. Daniëlle Bruggeman is a cultural theorist, specialized in fashion and identity. She is Professor of Fashion at ArtEZ University of the Arts in Arnhem, the Netherlands. She holds a PhD in Cultural Studies, which was part of the first large-scale interdisciplinary research project on fashion in the Netherlands, 'Dutch Fashion Identity in a Globalised World' (2010-2014). Daniëlle has been a visiting scholar at Parsons, the New School for Design (NYC), and at the London College of Fashion.

Jan Mahy, *Professor smart functional materials & innovative textile development at Saxion University of Applied Sciences*. Jan Mahy worked at the technical textile firm Low&Bonar and takes this expertise further in the research and education field via his professorship Smart Functional Materials & Innovative Textile Development at Saxion University of Applied Sciences, Enschede, the Netherlands.

Rens Tap, *Business strategist at Modint*. Rens Tap has over 30 years of experience in market research, industry projects and innovation, and is one of the most knowledgeable people in the field of business development for clothing and textiles. He works as a business strategist at Modint, the Dutch association of manufacturers, importers, agents and wholesalers of clothing, fashion accessories, carpet and (interior) textiles. The 500+ member companies of Modint achieve a combined annual turnover of € 9 billion in the Netherlands, of which more than 50% is exported.

1
The first pillar is based on the inaugural lecture and publication: Bruggeman, D. (2018). *Dissolving the Ego of Fashion: Engaging with Human Matters.* ArtEZ Press.

2
Sobchack, V. (2004). *Carnal thoughts: Embodiment and moving image culture.* Univ of California Press.

3
Van den Eijnde, Jeroen & Bruggeman, Daniëlle (2017) 'Imagining the Future through Speculative Design: Towards a New Paradigm where Art meets Science', conference paper, Cumulus Letter to the Future, Bangalore, November 2017.

The digital age introduced the possibility of wearable tech to the consumer: The *ICD+ Jacket* launched by Levi's and Philips in 2000 had extra pockets for a mobile phone and an MP3 player, next to hidden fabric loops for the cables. Traditional technology (garments) is used to wrap the new.

The SKIN probe project, *Bubelle* 2006, examines the future integration of sensitive materials in the area of emotional sensing— the shift from 'intelligent' to 'sensitive' products and technologies. As part of SKIN, Philips Design developed a 'Soft Technology' outfit to identify the future for high tech materials and Electronic Textile Development in the areas of skin and sensing emotion. The dress show emotive technology and how the body and the near environment can use pattern and color change to interact and predict the wearer's emotional state. Design Team: Clive van Heerden, Jack Mama, Sita Fischer, Rachel Wingfield, Stijn Ossevoort, Lucy Mcrae, Nancy Tilbury and Matthias Gmachl.

The aim of the *Lumalive* project was to light up T-shirts with dynamic light patterns, by integrating LED as smoothly as possible into garments. Philips worked on the development of fabrics that could digitally emit colors, which would potentially open a complete new way of generating fashion. The project brought forward shirts, dresses and furniture. The prototypes and products shown here are made with the help of by-wire.net.

vhmdesignfutures.com/project/192/
youtu.be/t5h_pGnL5l0
youtu.be/2l8jpZQkORc

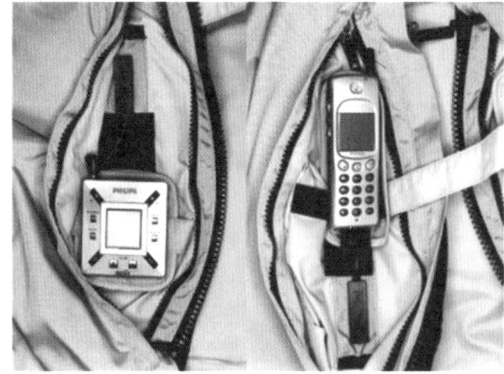

Philips/Levi's ICD+ Jacket

Levi's ICD+ Jacket
Bubelle
Lumalive

Philips

Philips *Bubelle*

Philips *Lumalive*

2000-09_01

Marina's master research (MA at the University of the Arts Utrecht, Fashion Design) included projects with emerging technologies, innovative textiles and new production techniques in collaboration with partners in the industry. This resulted in six garments that offered a look into the future of fashion

Triangled Coat is a fully customized coat inspired on 3D body measuring technology developed by TNO. The measurements are used to transform this in both the triangle-shaped pattern parts and the decorative print. The coat was digitally printed at the University of the Arts Utrecht, and fixated by Print Unlimited. This showed that it is possible to create fully customizable garments via digital production tools, based on one's own unique body measurements.

Huggy Care is an antibacterial dress created for Kwintet KLM, the aim of which was to inspire for less-clinical looking care wear that is still highly functional by adding an antibacterial function. *Huggy Care* is sweater-dress made out of synthetic yarn, combined with silver yarn that features nanotechnology, which gives antibacterial properties.

The *Tecatud Coat* was created with Ten Cate reflective hydro control textiles, in collaboration with Delft University of Technology. The reflective fabric, created with tiny glass balls, is optimized to provide better visibility for safety in traffic. It also features applied electronics for better user comfort.

by-wire.net/master-triangled-coat/
by-wire.net/master-huggy-care/
by-wire.net/master-tecatud/

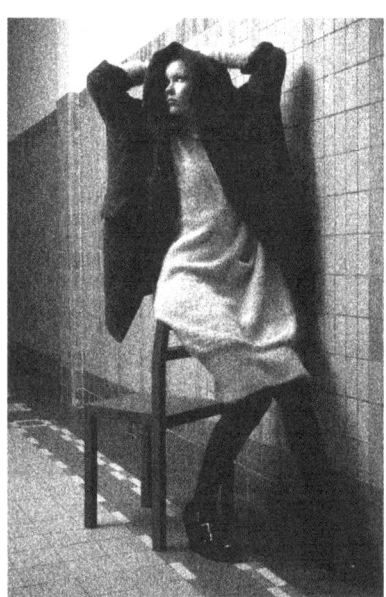

Triangled Coat, photo by Bob van Rooijen, model Sarah Nuiver
Huggy Care, photo by Bob van Rooijen, model Sarah Nuiver

Master Projects:
Triangled Coat
Huggy Care
Tecatud Coat

by-wire.net,
Marina Toeters

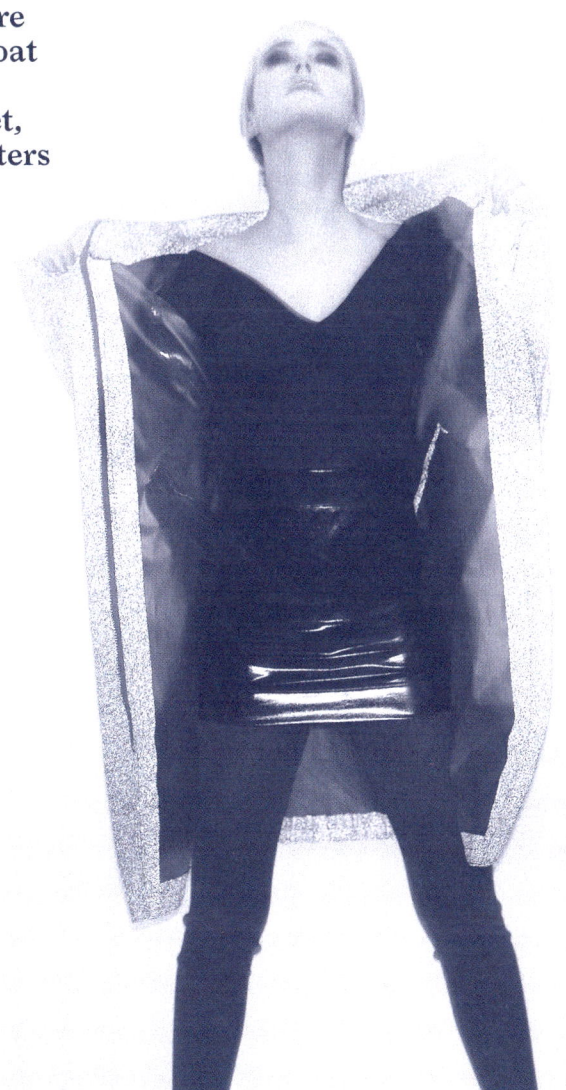

Tecatud Coat, photo by Bob van Rooijen, model Nina Wormer

2007_02

The *Fire Fighter Suit* is a bodysuit made by Ten Cate protective textiles for firefighters. The bodysuit can be worn underneath other clothing. While the textile itself regulates moisture, is breathable, as well as fire- and chemical resistant. The suit shows that it is possible to create garments out of fabrics that offer protection against multiple hazards.

Ecological Suited is an outfit of eco-friendly yak wool and hemp. Hemp fabric is five times stronger than cotton, as well as more durable. Additionally, the production process of this fabric is more environmentally friendly than that of cotton. The result was a fashionable, sporty outfit featuring workwear elements.

Dyna Seat Dress is made by 3D cutting Dynafoam into a wearable shape. The result was a dress-seat, a garment that can be used as dress, as well as seat. The inspiration for the design of the dress centered around the idea of 'he who looks at the past, will not get lost in the future'; thus, the shape was inspired by baroque fashion, contrasted with the innovative new Dynafoam material.

by-wire.net/master-fire-fighter-suit/
by-wire.net/master-ecological-suited/
by-wire.net/master-dyna-seat-dress/

Fire Fighter Suit, photo by Bob van Rooijen, model Jocelyne Norbruis

Master Projects:
Fire Fighter Suit
Ecological Suited
Dyna Seat Dress

by-wire.net,
Marina Toeters

2007_02

The Charlie project hacks the iconic Burberry coat to read fabric punchcards. When detecting a punchcard the coat plays the corresponding story from an old man's life through its headphones. The coat is part of *Media Vintage*, a series of interactive electronic textiles that contain memories. In *Media Vintage*, digital information is physically stored in textiles and read through interaction. The project is inspired by an article by Bruce Sterling in *The Book of Imaginary Media*, in which he points out that new media becomes obsolete faster than old media. The project is nostalgic for a time when technology was built to last and tries to imagine how digital data storage could be visual, physical and meaningful. It uses electronic textiles to suggests a potential future where digital data can be read by both man and machine.

Created with the support of the V2_ Institute of the Unstable Media with Piem Wirtz, Stan Wannet, Simon de Bakker, Joachim Rotteveel and Meg Grant.

melissacoleman.nl/media_vintage

Media Vintage–Charlie

Melissa Coleman

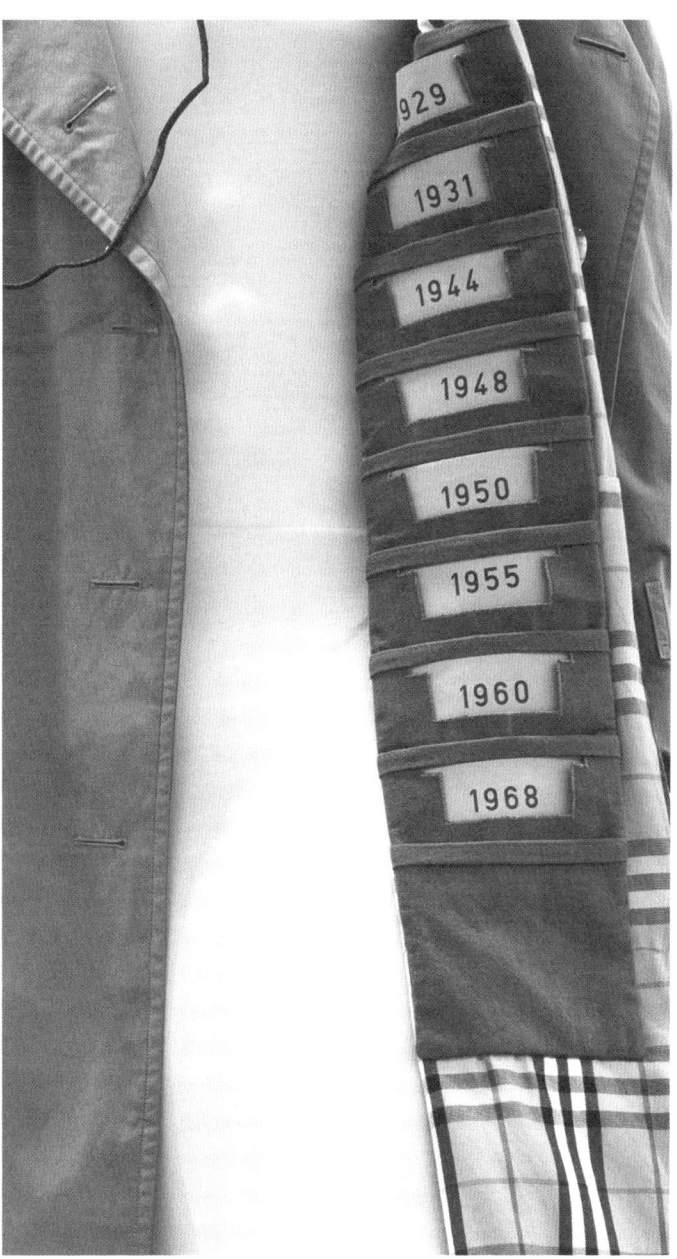

2009_03

A lot is happening within the development of textile technology, but the step towards application in fashion is sometimes too far. New textile technology first usually finds its way towards technical applications such as geotextiles instead of fashion and interior design. This collection serves as an example that there are technical possibilities for the use of new, already-available innovative man-made textiles within interior design and fashion textiles. The textiles used were 100% polyester to motivate an easy recycling process.

Thanks to Innofa, Philips lumalive, University of the Arts Utrecht, TextielMuseum, Protospace, Optima Knit, Igepa, Systemmag, Schoeller and Hero Textil.

by-wire.net/collaborative-textile/

Photos by Brian Smeulders, hair and visa Andrea Ligthart, model Mirte Kopinga

Collaborative Textile:
Fashion vs Interieur

by-wire.net,
Marina Toeters
JSSSJS,
Jesse Asjes

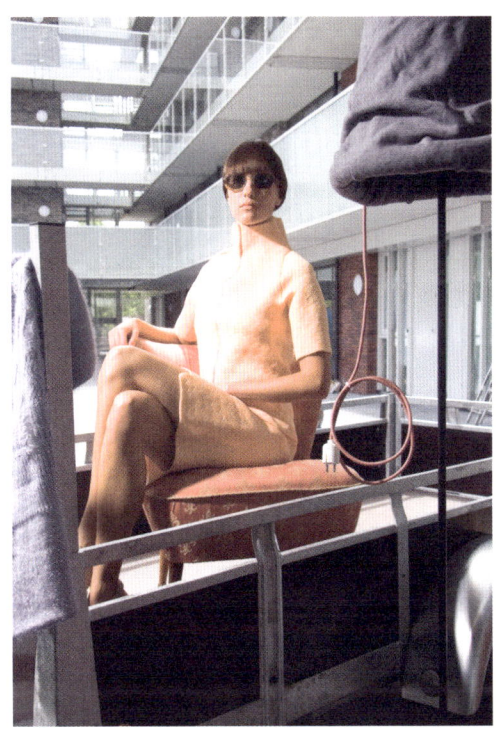

In the company of others people feel vulnerable when showing 'negative' emotions, such as grief, sadness, pain, fear, anger. *E-Pressed* wants to make space for these emotions and communicate them in a nonverbal way, by offering a different communication layer for handling the vulnerability caused by these negative emotions.

By sensing and visualizing inner states, *e-Pressed* creates awareness in the wearer and in others. Biometric sensors like Galvanic Skin Response (2009), respiration and muscle tension (2014) embedded in the shirt are indicator of the wearers stress level. Light visuals on areas originating from acupressure and trigger point therapy (2014) may provoke interaction and invite the wearer and others to press on them, relieving tension and stimulating well-being. In the second version the shirt provides both public and private touch and communication points enable users to choose the amount of exposure.

awarenesslab.nl/projects/e-pressed.html

Version 2009, Anja Hertenberger, Barbara Pais and Danielle Roberts: concept, design, research, interaction design, programming. Paul van Bavel: hardware.

E-Pressed: Promoting Awareness, Renewing Communication

Anja Hertenberger
Barbara Pais
Danielle Roberts

Version 2014. Anja Hertenberger, Danielle Roberts: concept, design, research, interaction design, programming. Rianne de Witte: fashion design, pattern, tailoring, laser cutting. Resi Kemperman: model. Funding: NBKS – Province of Noord Brabant, Tijlfonds

2009-14_05

chapter ii

By Ben Wubs

"Fashion is capitalism's favourite child, she sprung from its deepest being, and shows its specificity like hardly any other social phenomenon of our time", wrote the German economist Werner Sombart.[1] According to Sombart, fashion is an instrument that industry wields to mobilise consumption. It is a capitalistic phenomenon used by entrepreneurs to increase sales and profits. Fashion, which influences ever more sectors of economic life, constantly produces innovation, obsolescence, more innovation and further obsolescence. As a result, it has become one of the drivers of modern capitalism. Following this logic, fashion is at the heart of our economic system. Furthermore, to borrow the words of the Austrian-American economist Joseph Schumpeter (who didn't discuss the fashion industry in his work), one could say that fashion is part of, or even the foundation of the "perpetual process of creative destruction"[2]. In his view there are five types of innovation: new products or new qualities of products; new production methods; new forms of industrial organization; new markets; and new sources of supply. In the history of fashion innovation all five types can be identified. However, little attention has been paid, to the role that innovation and technology have played as a main driver of change in fashion.

The loom was one of the first innovations in human textile history. In pre-historical times, weaving and spinning were primarily female activities. The main fiber used in Mesopotamia (nowadays Iraq) was wool. Ancient Egyptians made clothes and other textile products based mainly on linen. Ancient Greece saw the use of both wool and linen, as well as the import of silk from China (beginning of the Silk Road) and cotton from India—underpinning the fact that textile production and markets had become global. The Romans introduced large scale fabric and dress production, mainly based on female slave labour. During the late Middle Ages silhouette and form became more important and male manual labour became more prevalent. Men began to weave and vertical looms were replaced by horizontal looms. With the rise of a rich urban bourgeoisie in North Italy and later North West Europe, craftsmanship in the cities became more and more important. Simultaneously, the organization of the textile industry was based on a 'putting-out system', meaning that parts of the production process took place in the homes of workers outside the cities.[3]

A major transition took place when steam power was introduced to the industry and mechanized looms increased productivity to unpreceded levels, first in Britain and later in Western Europe and North America. It revolutionized the production and consumption of textiles. Production increasingly took place in large factories in growing industrial agglomerations and cheap cotton fabrics became affordable for the urban lower classes. The industrial revolution was based on cotton production and is often portrayed as a Western success story, but consumption of cotton from Asia actually had created mass demand in Western Europe in the 18th century.[4] This in turn started a new plantation economy in North America, based on black slavery, and a global agro-industrial complex (Beckert 2014). A major innovation in the 19th century was the introduction of the mass produced and mass marketed sewing machine by the American businessman and inventor Isaac Singer. For the first time it was possible to set up large scale manufacturing of apparel. Total wars, such as the American civil war (1861-1865), created mass demand for uniforms and what came to be known as ready-to-wear clothing.[5] To this day, the global textile and fashion industry is based on these major innovations, albeit at a different, globalized scale. Because basic manufacturing principles are straightforward and transportation costs have plummeted, global supply chains have completely changed in the last five decades.

For a long time in human history, fabrics were made from 'natural' fibers, such as cotton, wool, silk and flax. However, in the 20th century, a synthetic revolution took place. A new industry grew up in Europe and North America in the 1920s: the artificial silk (rayon) industry. At the end of the 1930s, the American company DuPont introduced Nylon, the first man-made fiber based on petro chemicals. After World War II, many new synthetic fibres were developed. Polyester is the most famous example and is still the most dominant synthetic fiber globally. Around the year 2000, polyester claimed the top position from 'King Cotton', which had been the most commonly used fiber in the textile industry for 200 years.

However, synthetic fiber production is not without prob lem, as most synthetic fibers are not sustainable. But the solution is not to replace them with 'natural' materials. Cotton production, for example, has issues in relation to sustainability because it uses up large quantities of water and arable land that could be used for food production. Additionally, herbicides and pesticides are needed for the cotton monocultures. To make matters worse, growing numbers of fabrics use a mixture of synthetic and natural fibres, making recycling even more complicated.[6]

In conclusion, innovations in fashion have been the main drivers of change in the fashion industry, but has also created massive environmental issues which will not be easy to solve. The answers may be found partly at an industry level and may be technological in nature, like the digitalization of fashion and the creation of more sustainable fibres and textiles, but these issues need to be addressed at a transnational political level to cater for the interlinked sets of problems and the scale, scope and complexity of the issues.

Ben Wubs, *Professor International Business History, ESHCC Erasmus University Rotterdam*. Appointed Project Professor Graduate School of Economics, Kyoto University Japan. Ben Wubs is Professor International Business History at Erasmus University Rotterdam and engaged in various research projects related to multinationals, business systems, and the transnational fashion industry. Ben is writer of several books and teaches various subjects like Business History of Fashion, and International Relations Theory.

1
Sombart, W. (1902). *Wirtschaft und Mode. Ein Beitrag zur Theorie der modernen Bedarfsgestaltung Grenzfragen des Nerven- und Seelenlebens.* in Einzel-Darstellungen für Gebildete aller Stände. Zwölftes Heft, ed. L. Loewenfeld and H. Kurella Wiesbaden: J. F. Bergmann.

2
Schumpeter, J. A. (1943). *Capitalism, Socialism & Democracy.* London: Routledge.

3
Fortunati, L. (2010). Wearable Technology. In J.B. Eicher & P.G. Tortora (Eds.). Berg Encyclopedia of World Dress and Fashion: Global Perspectives (pp. 100–106). Oxford: Berg. Retrieved April 30 2019, from http://dx.doi.org.access.authkb.kb.nl/10.2752/BEWDF/EDch10013

4
Nierstrasz, Ch. (2015). *Rivalry for Trade in Tea and Textiles. The English and Dutch East India Companies (1700-1800).* Basingstoke: Palgrave Macmillan.

5
Tortora, P.G. (2010). Technology and Fashion. In P.G. Tortora (Ed.). Berg Encyclopedia of World Dress and Fashion: The United States and Canada. Oxford: Bloomsbury Academic. Retrieved April 30 2019, from http://dx.doi.org.access.authkb.kb.nl/10.2752/BEWDF/EDch3213

6
Blaszczyk, R. L., & Wubs, B. (2018). *The Fashion Forecasters: A Hidden History of Color and Trend Prediction* (4e ed.). New York, United States: Bloomsbury Academic.

The projects began by focusing on the standardised crochet lace 'how to' pattern book instructions, and translating them into computer code. The code created familiar crochet lace patterns on screen. As virtual lace the patterns were 'performed' as animations and showed how the patterns could be created stitch by stitch, and then 3D printed to create a lace object. Alternatively, the code could be disrupted using random variables in the code. The result was a series of emerging patterns that had never been seen before and speculative lace patterns that could not easily, if at all, be produced physically. When examining the coordinates of each stitch as the virtual patterns formed it became apparent that the some of the virtual patterns were over one kilometre wide and high. Some patterns created from this process initially radiated out and then imploded, while other patterns became 'locked' in a never ending repetitive process.

gailkenning.wordpress.com/art-design/evolutionary-lace/
videos.files.wordpress.com/quRXOZsW/showreel0_dvd.mp4

Pattern As Process

Pattern As Process, Evolutionary Lace, Evolving Lace

Gail Kenning

Digital visualisation *Evolutionary Lace*

2009-16_06

The *Human & Kind* project is created in collaboration with the European Space Agency. It explored the question of how textiles and garments can be used in the hypothetical case of living on the moon. This is researched not only from a technical perspective (can the materials protect from heat, cold, radiation?) but also if the textiles can make its wearer feel at home when being so far away. This was done by testing textiles on various characteristics, eventually choosing those which were best resistant to heat, the best conductor, the lightest material and the most social connected textile. The aim was to choose textiles that feel soft to their wearers and by finding applications that stimulate human kindness, such as a shirt that lights up when sensing life, or the appearance of a flower pattern when the environment becomes too hot for habitation.

Thanks to confection: katoenenzo.nl, Melissa Bonvie, lace factory: Museum de Kant Fabriek, electronics: Contrechoc, knitting: jsssjs.com, Jesse Asjes, laser/plot: Protospace, nitrogen: KI Dalfsen and technical support: Matthijs Vertooren

by-wire.net/moonlife/

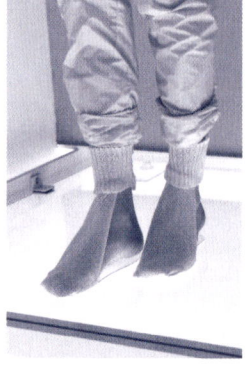

Human & Kind, photos by Maarten van der Meer, hair & visa Anna Edvardsen, models Melinda, Kyra, Adam, 77 models

Human & Kind

ESA
by-wire.net

2010_07

With this shirt the wearer can feel Beethoven move on their body, or feel Tiësto beating all over their arms and back. The shirt implements 64 carefully integrated tiny vibrators, all of which can be individually controlled. A microchip with wireless connection is connected to a system that translates music in tactile information, which is then picked up by the vibrators. In practise, high tones will be felt high upon the body, and low tones activate the vibrators situated lower within the shirt.

by-wire.net/100702/

Vibrating Shirt
Reacts to Music

University of the Arts Utrecht,
Tim Walther
by-wire.net,
Marina Toeters

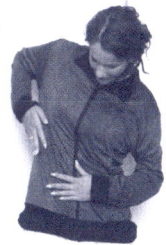

Photos of the harddware inside and during interaction
by Maarten van der Meer

2010_08

Since 2010 the developments towards textiles where supported by funding like the European project PLACE-it. Philips Research was highly involved in this. It made it possible to discuss and to test prototypes in the lab and with Philips commercial products in real life. by-wire.net contributed by developing some of the textile parts of the prototypes. What follows are some of the prototypes.

Philips *BlueTouch: Pain Relief Patch* 2010-2013, is an example of wearable technology that is already available on the (medical) market. *BlueTouch* is an array of high intensity blue light emitting LEDs pointing towards skin on the lower-back, in order to treat muscle pains. To create better wearability and eventually acceptation, by-wire.net developed the straps used to attach Pain Relief Patch onto the body, as well as the research and development, prototyping, user comfort studies, product design, material (textile) research, fitting/sizing and production sourcing. Here Philips has shown that medical devices use fashion where it comes to making the products soft and bendable. Users will appreciate and accept products better when they look good and familiar.

Blue Light Wristband 2013, was developed to treat wrist pain and uses the same technology as BlueTouch. The integrated technology is from Imec and Holst Centre.

Philips *BlueControl: Psoriasis therapy device* 2014, is used to treat psoriasis. Light therapy is one of the ways with which patients with this condition can find some relief. by-wire.net contributed to the development of this blue LED light therapy device, which is unobtrusive in its use, for use by skin therapists.

ec.europa.eu/digital-single-market/en/news/place-it-electronics-wear-light-health-care
by-wire.net/philips-blue-touch/
philips.nl/c-m-pe/pijnverlichting/blue-touch
by-wire.net/wristband-techtextil-2015/
by-wire.net/philips-blue-control/

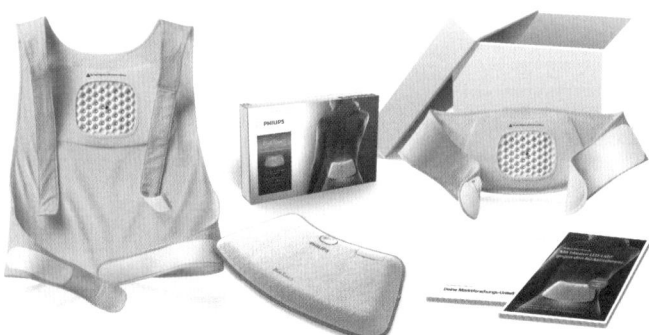

BlueTouch, Philips

Philips Medical Projects:
BlueTouch
Blue Light Wristband
BlueControl

Philips Research

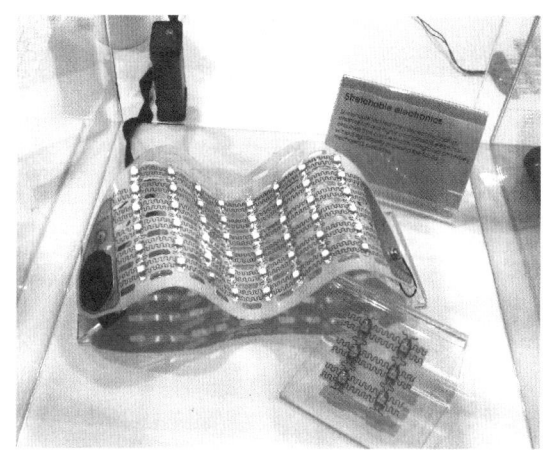

Blue Light Wristband, Philips

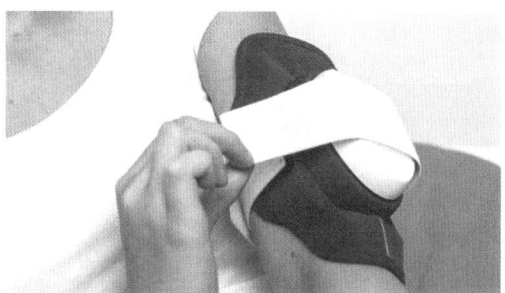

BlueControl, Philips

2010-14_09

Drapely-o-Lightment is a skirt based around the ideas of 'drape' and 'light'. The skirt is made up of 2500 triangle-shaped patches, while the lights are 6 OLEDS (organic light-emitting diodes), integrated within the fabric. The triangles create interesting shapes and drapes, which are accentuated by the lights.

Collaborators: Prof. Dr. Ir. Loe Feijs, TU/eindhoven and Koen van Os, Philips Research.

by-wire.net/20121123/
mitpressjournals.org/doi/abs/10.1162/LEON_a_00913 /
youtu.be/mgjJz_HMU1s

Drapely-o-Lightment

Laurentius Lab
Philips Research
by-wire.net

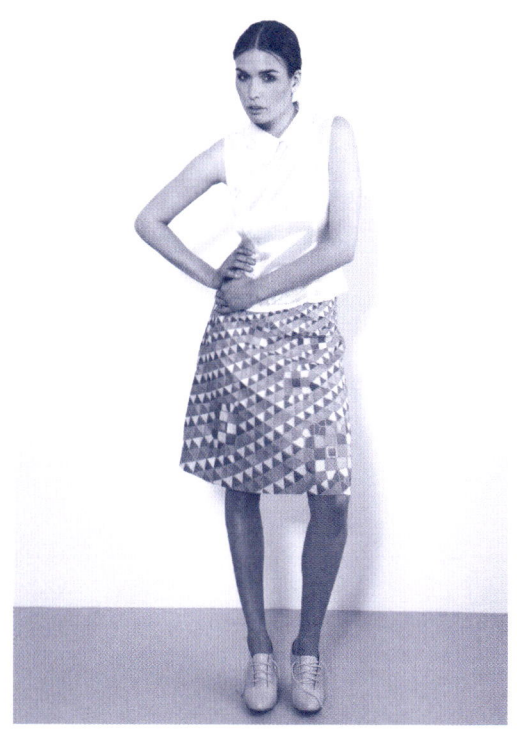

Photos by Brian Smeulders, model Stephanie Samson

2012_10

On the eve of the 80th anniversary celebrations of the brand, Lacoste pays tribute to René Lacoste and his visionary spirit by imagining a future version of the brand's most iconic creation, the polo. The brand asked to come up with an international digital campaign.

Lacoste presents a clip in which the emblematic clothing piece is projected into a future where textile technology knows no boundaries. The clip comes with an invitation towards fans to create their own story featuring the Polo shirt's futuristic image.

mnstr.com/en/work/lacoste/
vimeo.com/77200317

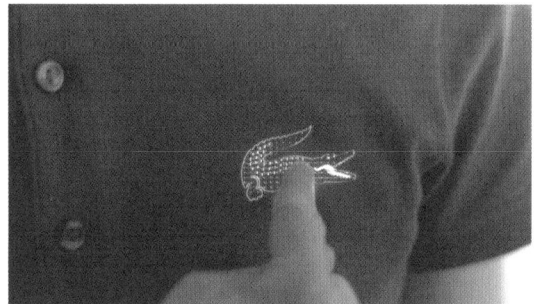

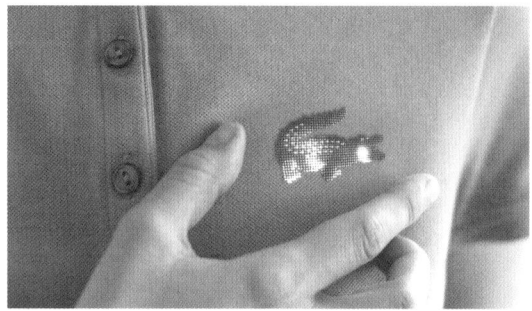

Polo of the Future

m.nster. and
Studio Roosegaarde
for Lacoste

2012_11

chapter iii

By Loe Feijs and Koen van Os

Technology and its promise

Loe: Nano-coatings, electric circuits, energy-harvesting, biometric sensors, smart protection: there is no lack of promising wearable technologies. But which technologies will make a significant change to garments, the way we use them and how they connect us to the world? Garment construction has always been technology-dependent. Companies such as Singer and Brother filed thousands of patents since Singer's first successful sewing machine patent in 1851. The most significant technological developments of recent years have been in information technology. The digital age introduced the possibility of wearable tech for the consumer. For example, the ICD+ jacket launched by Levi's and Philips in 2000_01 had extra pockets for a mobile phone and an MP3 player, with hidden fabric loops for cables. Traditional technology is used to wrap the new.

2006-09_01

Koen: Philips was at the forefront of the early implementation of electronics in textiles. During the Lumalive project 2006-09_01 we experimented with the concept of wearable tech in a project open to the public. There was a great risk of failure, but there were also many lessons to be learned. What followed from 2009 on was very much based on the experiences gained during Lumalive. As former engineers we had to get used to a new way of doing research as alongside hard science and technology, 'soft' opinions began to influence the course of new Philips products.

The exciting world of textiles

Koen: The Lumalive project opened up a world of possibilities, new business opportunities, new markets and new expectations. The public expected smooth integration and robustness in electronic textiles and wanted products that can be laundered in a washing machine. After all, if Philips could develops light-emitting fabrics, then it must also be possible to have all of the other material qualities expected of textiles. At the end of the Lumalive project, I moved from being an engineer to an industrial scientist in the Philips Research department. It was the perfect place to engage

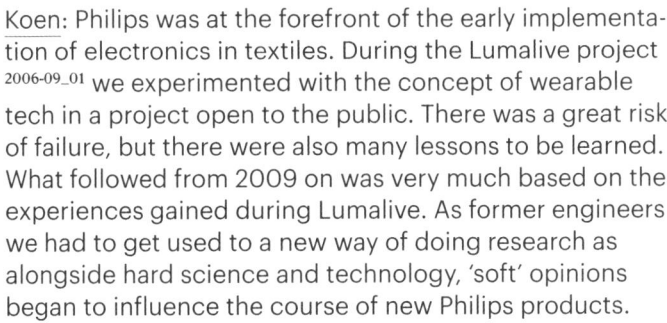

2014_09 BlueControl, Philips

2006-09_01 Lumalive Project, Philips

2010_08 Vibrating shirt reacts to music, HKU and by-wire.net

2016_32 Bambi Medical

in experimentation and discussion with peers. Philips was central to discussions about e-textiles and wearable electronics, and with the support of subsidy funded projects like the European project PLACE-it [2010_09] it was possible to engage with the industry, test out ideas in the lab, and release commercial products to a wider public.[1]

Medical applications

Koen: Two new product categories emerged from the Lumalive and PLACE-it projects. The first one was the professional LED lighting business. Lumalive showed how fabrics could be digitally coloured and programmable. This concept was later reused and scaled up for a metre wide wall-panel used for interior design. The second category development was in medical products such as Philips BlueTouch [2013_09], BlueControl [2014_09] and Smart Sleep [2017_40] which collected knowledge about the wearability and battery use in smart garments.

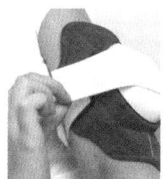

2014_09

Loe: In the university there was also a lot of focus on medical products close to the body. Through the invention of conductive yarns, electric wiring becomes flexible and soft. The garment becomes a circuit, a sensor, an actuator [2010_08], or all of them. This offered possibilities for knitted ECG (electrocardiograms), respiration sensors, stretch sensors [2015_25] and embroidered vibration-motor coils [2016_14]. Here at the University of Technology Eindhoven (TU/e), I am pleased we had the opportunity to work with a neonatal department in a hospital to make a soft-sensor baby jacket.[2] This was later commercialized by Bambi Medical [2016_32].

2010_08

Advent of the smartphone

Koen: We had assumed that the smooth integration of electronics into wearables could only be realised when electronics could be woven or knitted at smallest level possible. However, this was not the case, and around 2014 we saw the advent of smartphone apps and wristwatch designs become wearable, without using textiles at all.

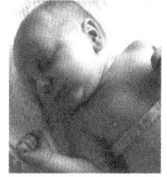

2016_32

Loe: Unfortunately, the adoption of technology in wearable products for the consumer markets is disappointingly slow, especially when compared to the expectations raised around 2000. For technologists, like myself, it was hard to see the full potential the smartphone would have. But, the smartphone carries its energy storage, has a user interface and has powerful sensors, actuators, networking capability, and is light enough to be carried or 'worn' in garments. Right now, most smart garment innovation is happening in the domain of high-performance sports and the medical domain, and the findings will likely trickle down to the consumer markets, eventually.

Lessons learned

Koen: We experienced a range of issues, with the integration of conductive electronics into textiles. For example, lifetime disbalance. An electronic device needs to continue working after twenty washes, otherwise guarantees of replacements are needed. Whereas, with other textile products users accept, for example, the fact that the colour black will fade, gradually. Another point of discussion was the end-of-life treatment of smart products. Bringing together electronics and textiles massively complicated an issue that needs to be resolved before any type of mass-market implementation can happen. Practical issues arose, such as the differing discourse used in different industries which made the industrialization of concepts challenging. For example, the textile partners preferred to produce large quantities of material. Whereas we, as electronic manufacturers, only needed a few meters of fabric for our first batches of functional products.

Digitalization in manufacturing

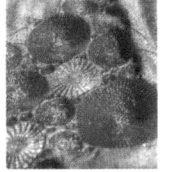

2016_14

Loe: The machines and the processes of making garments are changing. Designers are experimenting with 3D printers and laser cutters.[3] Whereas a typical machine in a traditional factory has a complicated set-up procedure, after which it can produce high volumes at high speed, these new machines are set up from a single file and can create individual pieces. The same holds for digital embroidery [4] [2016_14], digital printing [2015-18_30], and digital weaving[5] [2017_14]. Most machines are heading in the direction of enabling fast diversification and even personalization.

Koen: In the digital era, industries have had to re-shape their organisation. At Philips Lighting, now 'Signify', we are experimenting with 3D printing. At first glance, 3D printing seems like a textile method. Machines are replicated, like looms in a weaving factory, processes are slow, and everything starts with big spools of material. But, the process is completely digital, and the materials are circular.

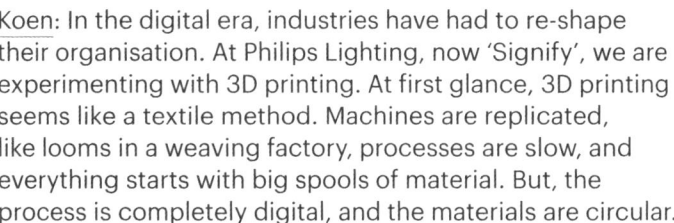

2016_14 Pied de Pulse, Laurentius Lab, Loe Feijs and by-wire.net, Marina Toeters

1
European project
PLACE-it (2014,
June 27). PLACE-IT:
Electronics-to-wear
light up health care.
*Digital Single Market,
Projects story* https://
ec.europa.eu/digital-
single-market/en/news/
place-it-electronics-
wear-light-health-care

2
Bouwstra, S. (2013).
*Designing for the
parent-to-infant
bonding experience*
Eindhoven: Technische
Universiteit Eindhoven
DOI: 10.6100/IR760049

3
Feijs, Loe, and Marina
Toeters. (2015).
"Drapely-o-lightment:
An algorithmic
approach to designing
for drapability in an
e-textile garment."
Leonardo 48, no. 3
(2015): 226-234.

4
Feijs, L. M. G., and
Marina Toeters. (2016)
"Pied de pulse: packing
embroidered circles
and coil actuators in
pied de poule (hound-
stooth)." Proceedings
of Bridges: 415-418.

5
Feijs, L. and Toeters,
M., (2018). Cellular
automata-based
generative design
of Pied-de-poule
patterns using
emergent behavior:
Case study of how
fashion pieces can
help to understand
modern complexity.
International Journal
of Design, 12(3),
pp.127-144.

Every minute a new product can be started. A designer can create a new design and fully materialize it within the hour, without needing the help of colleagues. The 3D printed products begin solid, but gradually become soft. The holy grail (for me!), of making 3D printed digitally coloured fabrics [2012_10], [2015_20], [2016_33], [2018_41] could be within reach in the coming years [2017_37]. We might be moving towards a digital approach which echos the virtual digital dynamic changing fabrics in fashion seen in Angella Mackey's work [2019_50].

Fitting into systems

Loe: Currently, each garment is connected to another and exists within a system. A piece of clothing does not stand on its own, it is part of a brand's collection, or part of today's outfit. A smart garment with, for example, an embedded sensor [2009-14_05], [2015_25], needs to offload its data and connect to another object. If it is an energy harvester [2012_13], it may share its power with other wearables or other clothing pieces. Therefore, it is not only the garments that are connected. The companies that produce them, the end-users and the young designers are all digitally connected too. The call for sustainability can no longer be ignored. If technical universities and the fashion industry collaborate we will begin to see real innovation in the fashion system.

Professor Loe Feijs, *Vice Dean Industrial Design, Eindhoven University of Technology.* Loe Feijs has an MSc in electrical engineering and a Ph.D. in computer science. In the 1980s he joined Philips working on telecommunication systems and fundamentals of software design. At present Loe is professor for Industrial Design of Embedded Systems. Loe is the author of three books on formal methods [6] and of over 100 scientific papers. We collaborate at Eindhoven University of Technology education students in the Wearable Senses Lab where I work as Industry Liaison. We work on the crossing of mathematical principles and fashion. We try to advance pattern design and production methodologies for garments [2012_10, 2013_14].

Koen van Os, *Wearable Technology and textile expert at Philips Lighting.* Koen van Os has a MSc in mechanical engineering and further educated in the rich technology environment of Philips. Within Philips departments (today at Signify) he worked on many products and manufacturing technologies like LEDs as new light sources for consumer use. In 2006 Koen got formally involved in e-textiles via the Philips Lumalive project [2008-09_0] and became specialized in various ways of manufacturing electronic textiles and making market ready wearable tech products.

The Waste Conscious Scarf measures the air quality around the scarf. However, this was not the project's main objective: This scarf was made to increase awareness of the problems created by textiles containing electronics. Waste was measured during production. While fashion tech is proposed as an alternative to fast fashion, the use of electronics could potentially create an even larger heap of waste if the production follows the same process as the fashion industry does today. Designers should be able to produce a design which takes recycling and pollution into account before e-textiles enter mass production.

by-wire.net/20121120/

The Waste Conscious Scarf

Contre Choc
by-wire.net

Photos by Brian Smeulders, model Stephanie Samson

2012_12

The idea behind *Solar Fiber* is that of a flexible photovoltaic fiber that converts sunlight energy into electrical energy. The aim was to develop this as a yarn that can be worked into all sorts of fabrics. This 'smart material' can be used in all types of applications where textiles are currently used, but with the added advantage of being able to produce an electrical current. Developing a photovoltaic fiber is not a completely new idea, but nor is it a simple task. The approach is two-fold: First and foremost, *Solar Fiber* is working on a photovoltaic fiber with a protective coating that will likely start its life as a 5mm fiber and eventually be extruded to 100 m. Before it gets there, *Solar Fiber* is working on proof of concept prototypes that will help to communicate our idea and show real life applications for the technology. Three prototypes show this approach.

The *Solar Jacket* 2012 used rigid solar cells to see how feasible their application would be at that point in time. While solar cells are not easy to integrate in textiles and not comfortable, it was possible to charge a battery and charge a phone using the energy collected by the solar cells.

The *Solar Fiber Shawl* 2013 woven by Van den Acker Textielfabrieken shows that it is possible to weave optical fibers in an industrial manner.

The *Solar Fiber Knitted Shirt* 2015 continues the experiment. Optical fibers are integrated during the knitting process by Jesse Asjes. Tiny photodiodes are connected to the end of the fibers to transform the transported light into electrical current. It proved that the energy from solar fibers can be harnessed.

by-wire.net/solar-fiber-jacket/
by-wire.net/solar-fiber-knitted-shirt-with-jsssjs/
solarfiber.nl/

Solar Jacket, photo by Brian Smeulders, model Stephanie Samson

Solar Fiber

**Meg Grant
Ralf Jacobs
Marina Toeters
Aniela Hoitink**

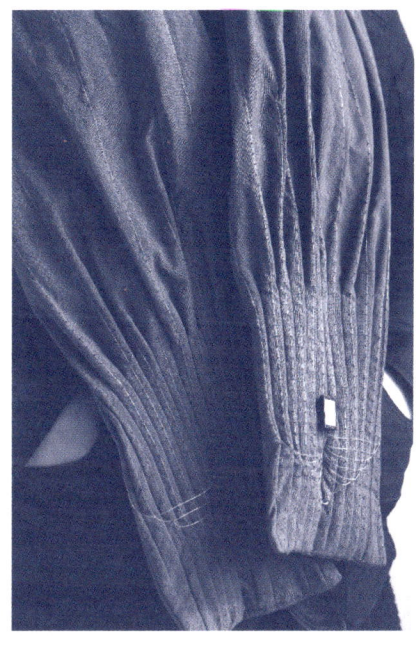

2012-15_13

Pied-de-poule is a classic textile pattern, with the use of a mathematical theory we generate new and interesting Pied-de-Poule patterns. With these four projects we showed that mathematics and fashion work in great collaboration.

fractalPDP: a novel textile pattern 2013 is the first explorations of using mathematics to generate new patterns for fashion design. The inspiration comes from the mathematical Cantor set theory and the final pattern was generated by running a recursive algorithm in Processing. The outer layer of white fabric has been laser-cut and the tiny holes reveal the black layer underneath. When you see the garment from far you will see a big houndstooth, and when viewing the garments closer you will see it is made of smaller pied-de-poule. This is called a fractal.

In this *Line Fractal Pied de Poule* 2015, the fractal is used again but now the pied-de-poule is implemented as a single line. Instead of blocks which get fragmented into smaller and smaller shapes, it begins with a single continuous line which is expanded by adding more and more nested zigzags.

by-wire.net/fpdp-2/
instagram.com/laurentiuslab/
by-wire.net/line-fractal-pied-de-poule/

Left: fractalPDP, photo by Brian Smeulders, model Stephanie Samson

Pied-de-Poule (PDP) or Houndstooth Projects

Laurentius Lab
Loe Feijs
by-wire.net,
Marina Toeters

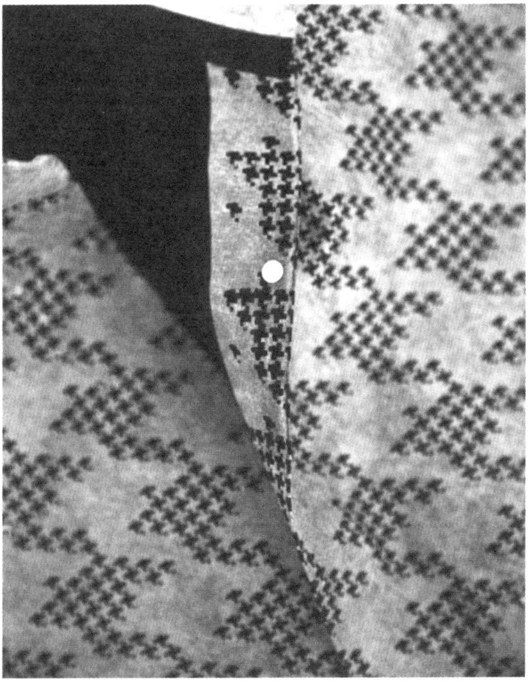

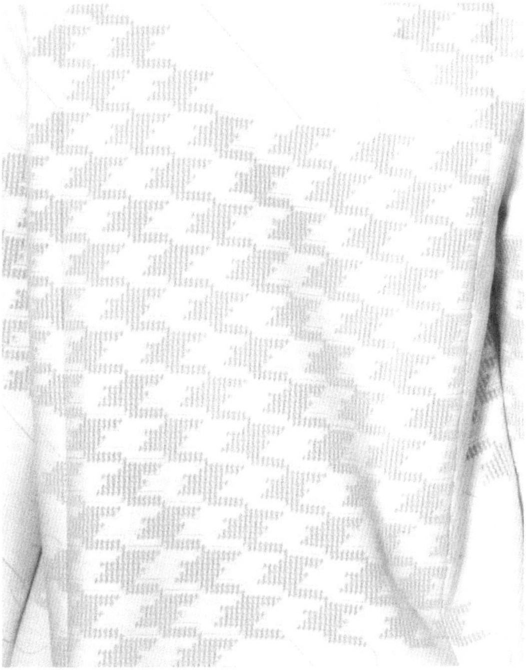

Line Fractal Pied de Poule, photos by Brian Smeulders, model Renata van Putten

2013-17_14

Pied de Pulse 2016 has two aims: The first is to study and implement a fractal-like structure of circles inspired by Apollonian circles, combined with a pied de poule. The second aim is to push the integration of electric actuators in garments, using the power of algorithmic design and digital manufacturing. Flat coils of copper with magnets work as vibration actuators in the garment.

A Cellular Automaton for *Pied-de-poule* 2017. With the use of the cellular automata theory a more abstract pied-de-poule can be generated. Because of the complexity only the smaller version, a so-called puppytooth pattern is used. The generator only works with five colours. The desired pattern appears as a contrast of darker and lighter colors, resembling the traditional black and white pattern of pied-de-poule. Sometimes the puppytooth is clearly visible while in other places it is more scattered, this creates a dynamic design. The resulting pattern was used to weave fabric by Van Engelen & Evers, and made into a small collection.

by-wire.net/pied-de-pulse
youtu.be/-LDk4T3y-1M
by-wire.net/cellular-automaton-pdp/
youtu.be/e2Zir3UcUwc

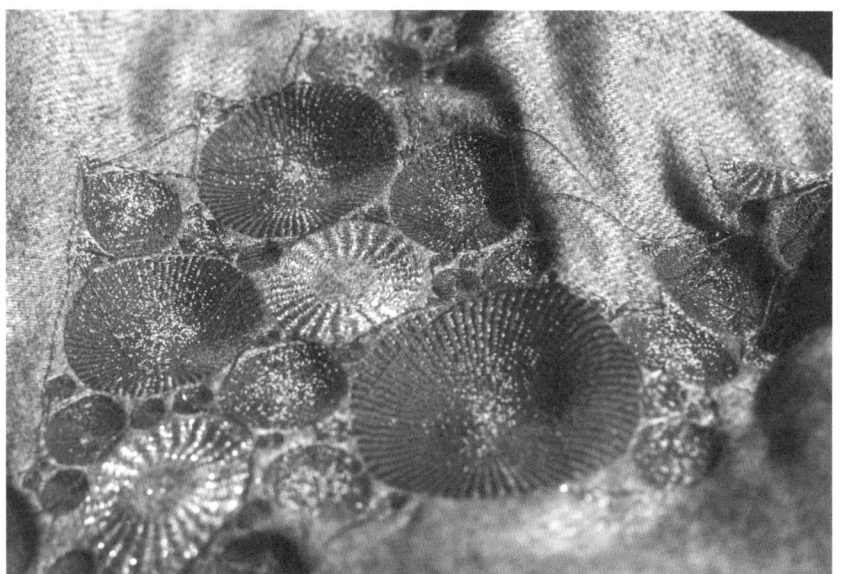

Pied de Pulse detail, photo by Flora Macleod

Pied-de-Poule (PDP) or Houndstooth Projects

Laurentius Lab
Loe Feijs
by-wire.net,
Marina Toeters

Cellular Automaton for Pied-de-poule, photos by Robin van der Schaft, styling by Maaike Staal

2013-17_14

Smart Textiles Services opens up a vast field of opportunities for textile developers and product and service designers. This project explored how the creative industries could help the textile industry to move from product to services, from closed to open innovation, and from vertical to horizontal production structures. In order to do so, the Wearable Senses Lab became a laboratory, enabling textile developers to understand the multi-disciplinary opportunities and challenges of creating Smart Textile. We selected three sample projects to illustrate the wider project.

Vigour 2013 by Martijn ten Bhömer and Pauline van Dongen is a knitted cardigan that keeps people active. Vigour enables geriatric patients, physiotherapists and family to gain more insight into the exercises and progress of a rehabilitation process. The cardigan has integrated stretch sensors made of conductive yarn and an app which monitors the movements of the upper body and can give sound feedback. The cardigan is knitted at Textiellab, Tilburg.

Vibe-ing 2013 by Eunjeong Jeon, Kristi Kuusk, Martijn ten Bhömer and Jesse Asjes is a self-care tool, which invites the body to feel, move, and heal through vibration therapy. The merino wool garment contains knitted pockets, embedded with electronics that enable the garment to sense touch and vibrate specific pressure points on the body. Knitted at Textiellab, Tilburg.

Spine Dress 2014 by by-wire.net, Lantor and Saxion is slightly warming your spine towards a comfortable temperature in tiny pulses. The conductive non-woven material of Lantor has a high resistance, the copper ribbons a very low resistance, and they spread the current vertically while the non-woven in the middle warms up when current passes through.

Contributors: De Wever, Metatronics, Unit040, Savo BV, TextielMuseum TextielLab, Ralf Jacobs and Contre Choc.

by-wire.net/smart-textile-services/
selemca.camera-vu.nl/projects/smart-textile-services.html
mtbhomer.com/
vimeo.com/101247686
kristikuusk.com
vimeo.com/user13422491/vibe-ing
by-wire.net/crisp-spine-dress/

Vibe-ing, photo by Wetzer & Berends

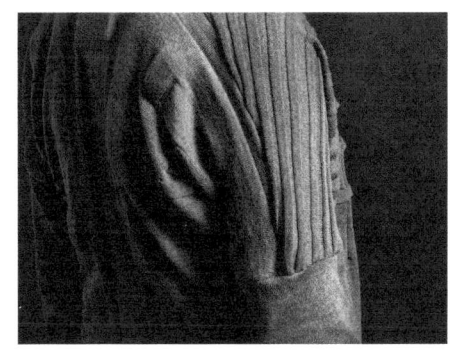

CRISP
Smart Textile Services

TU/e Wearable
Senses Lab
and many others

Vigour, photos by JR Hammond, Hammondimages

Detail Spine Dress by Wetzer & Berends

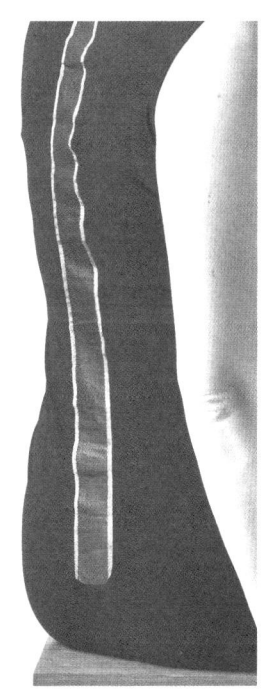

The *Solar Shirt* was developed in collaboration with Holst Centre. The solar shirt's main specification is in its innovative and progressive integration of solar cells into textile. Functioning as a fashion garment as an embodied interface, the shirt combines solar panels and flexible electronics.

The shirt seamlessly incorporates 120 thin film solar cells combined into standardised, functional modules using Holst Centre's stretchable interconnect technology for integrating electronics into fabrics. The *Solar Shirt* is designed as everyday wear that can charge a smartphone or any other USB compatible, portable device. With a fashion-forward design that includes practicality to be worn on a daily basis, studio Pauline van Dongen aims to take the solar fashion from the catwalk to the high street.

paulinevandongen.nl/project/wearable-solar-shirt/
vimeo.com/156573547

Photos by JR Hammond, Hammondimages

Solar Shirt

Pauline van Dongen

2014_16

MVO Caring workwear is targeted towards nurses and caregivers. by-wire.net created comfortable garments designed as uniform systems: The undergarment provides health professionals—who perform a very physical job—support to the shoulders, lower back and knees by pattern construction and choice of fabric. The outerwear is equipped with an anti-bacterial coating, which helps reduce the risk of bacterial contamination. Electronic 'wearables' in the garments can issue a warning signal in case of overload and unbalanced postures; wearers can track their posture behaviour with an app. A gas sensor warns for harmful substances in the air around the wearer.

The clothing is also friendly to its creators and the environment. Workers with a fair wage and in safe workplaces produce the pieces in factories in Tunisia and Bangladesh. The production is as environmentally friendly as possible, for example with C2C certified PLA/Tencel. Lastly, it can be recycled, which helps to reduce the waste of healthcare institutions.

Collaborators: Alcon Advies, BrabantZorg, by-wire.net, Dutch-Spirit, JJH Textiles, Newasco, Radboudumc, UMC Groningen, UMC Utrecht en Van Puijenbroek Textiel, process management by MVO Netwerk Zorg, in collaboration with the international experts of MVO Nederland.

by-wire.net/sustainable-and-supportive-garments-for-nurses/

Photos by Jan Willem Groen

Sustainable and Supportive Garments for Nurses

MVO consortium including by-wire.net

Sensor detail at the back and smartphone application

2014_17

chapter iv

By Lianne Toussaint

We are constantly touching and being touched by clothing and technology. This is why we so easily take their presence for granted and overlook what they mean to us. But what if fashion and technology—two 'things' that already constantly and closely surround us—became one?

The field of wearable technology helps us anticipate a future scenario in which clothing and technology are no longer separate domains. In a few years from now, I may no longer have to carry many technological devices, to press their buttons and touch their screens. Soon I may simply wear clothes to charge devices [2012-15_13], [2014_16], [2017_39] warm my spine [2014_15], correct my posture [2014_17] alert me to bad air quality [2012_12], [2015_25] or tell my surroundings how I feel [2009-14_05], [2015_22]. These innovative designs allow us to imagine ways of *living with*, rather than having our lives being impacted *by*, the technologies surrounding us. Exploring alternative possibilities for the use and 'wearing' of technology, fashion tech designers sketch a future scenario in which our interactions with clothing and technology are different and perhaps even more meaningful than ever before.

2012-15_13

The challenges of integration

Since the beginning of the twenty-first century, many scholars have predicted that the integration of fashion and technology will have significant cultural and societal impact. They have emphasized the expressive and creative potential of 'fashionable technology' [1], and argued that wearables will "have profound implications for our experience of body and mind, our communication abilities, healthcare and lifestyle". [2] Some even believe that wearable technology will radically disrupt the established fashion industry, ultimately displacing it for an entirely new market and value network. [3] One of the main barriers to success, however, is our still limited understanding of the possibilities offered by, long-term impact of, and day-to-day experi-

2015_22

2012-15_13 Soler Fiber, Meg Grant, Ralf Jacobs, Merina Toeters and Aniela Hoitink

2015_22 Fashion on Brainwaves, Jasna Rok

ences of wearable technologies. While it remains unclear what kinds of experiences, relations, and forms of communication these designs can bring about, both makers and potential consumers will remain ill-prepared for their definitive breakthrough. What does it mean to wear technology in different environments, different social settings, different cultures and at different times of the day or year? To answer these pressing questions and fully understand what wearable can mean for us, it is important to do more longitudinal, large-scale and qualitative research on how wearers experience these designs in an everyday context.

My Ph.D. research [4] took the first step towards a better understanding of the socio-cultural implications of wearing technology. I visited several fashion tech exhibitions and events, and spoke to several models and (test) wearers, exploring what fashion tech does to people and their behavior. The experience of wearable technology is a deeply embodied experience, that requires a "reinventing [of] our relationships to our bodies, our experiences of spaces, social interactions, and self-representation" [5]. I was struck by how both wearers and spectators interacted with the wearable technology, often responding with a mix of fascination, awe, and reservation. When researching the future potential and impact of wearable technology, it is important to remember that the extent and intensity of any effect varies from design to design, between different wearers, and in different contexts. A first step towards the successful integration of technical textiles, therefore, is to recognize the socio-cultural dimensions of wearing technology directly on, and in constant touch with the body. The different meanings and interpretations that fashion tech evokes in different wearers and environments, reveals that experiences of wearable technology should always be understood as socially and culturally situated.

The second challenge to overcome is understanding how the integration of technology impacts the dynamics of social interaction and self-presentation through fashion. Outputs, in the form of text, light, colour and movement, allows techno-fashion to express something about the wearer's personality, physical or psychological well-being, sports performance, or mood in unprecedented ways. Wearable technology can therefore, not only subtly or radically transform how wearers perceive themselves and communicate with the world around them [6], but also have profound effects on how others perceive the wearer. When we wear, rather than just carry or use technology, it becomes imbued with 'fashion aspects' such as social visibility, identity and self-expression [4].

The final step to be undertaken is to recognize that wearable technology, precisely because of its physical proximity to the body, can have both highly desirable and undesirable effects. Ethical issues such as privacy, power relations and autonomy should concern both developers and consumers. On the one hand, fashion innovations can help us to protect and care for ourselves or others in more conscious and effective ways. The fashion of the future may improve our physical experiences (a better posture, a warmer spine, better sports performance) and mental state (feeling less stressed, more comfortable, empowered, healthier, safer). However, wearable forms of surveillance and bio-monitoring risk compromising the well-being and autonomy of the wearer, especially when it concerns vulnerable or disadvantaged groups such as geriatric patients [2013_15], [2015-18_30] or children with autism [2016_31]. We need to be careful not to fall into the trap of technology push, where we are desperately looking for societal problems that we think can be solved by designing yet another new prototype. The ultimate challenge will be to carry out longitudinal research and design projects, co-initiated and critically assessed by those whose lives they will eventually affect.

2016_31

2016_31 Agent Unicorn, Anouk Wipprecht

Dr. Lianne Toussaint, *PhD
Researcher in Cultural Studies,
Radboud University, Nijmegen.*
Lianne Toussaint was a lecturer
and researcher at the department
of Cultural Studies of the Radboud
University Nijmegen, the Netherlands, where she taught courses
in the BA and MA programmes of
Arts and Culture Studies, including
Working through Fashion, Thinking
through Fashion and The Body in
the Arts and Visual Culture. Her PhD
research, funded by NWO, was about
the properties, design, and application of fashionable technologies.
From September 2019 she will be
teaching at the Department of Media
and Culture Studies at Utrecht
University, the Netherlands.

1
Seymour, Sabine
(2009), *Fashionable
Technology. The
Intersection of Design,
Fashion, Science, and
Technology,* Vienna:
Springer

2
Quinn, Bradley (2002).
Techno Fashion.
Bloomsbury Academic.

3
Disrupt Fashion
(2012-2016),
#DisruptFashion
Hackathon, San
Francisco.

4
Toussaint, L. (2018).
*Wearing Technology:
When Fashion and
Technology Entwine*
(Doctoral dissertation,
Radboud University
Nijmegen.

5
Lamontagne, Valérie
(2017). *Performative
Wearables: Bodies,
Fashion and
Technology.* PhD
Thesis, Concordia
University, Montreal,
Canada.

6
Smelik, Anneke (2017),
'Cybercouture: The
Fashionable Technology
of Pauline van Dongen,
Iris Van Herpen and
Bart Hess'. In: *From
Delft Blue to Denim
Blue: Contemporary
Dutch Fashion*, London:
I.B. Tauris, pp. 252-269

Bacteria Patterns are customizable fabrics. By growing the bacteria of our everyday life in petri-dishes and transfer printing it onto textile, a unique world can be captured on fabric. These fabrics are a result of a journey into the world of biotechnology. Biotechnology is predicted to be the technology of the 21st century. Tamara Hoogeweegen envision it coming together with DIY-movements and local economies to democratize the technology and empower society. Thanks to Pieter van Boheemen for making the Biohack Academy possible and Maria Botó for being supportive in the lab.

tamarahoogeweegen.com/
vimeo.com/130821080

Bacteria transfer printed

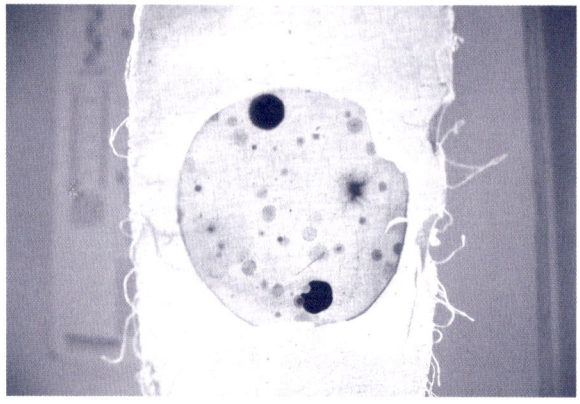

Bacteria patterns

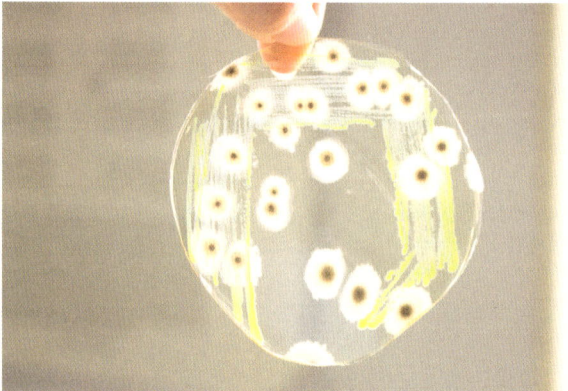

Posterus Textilus

Tamara Hoogeweegen

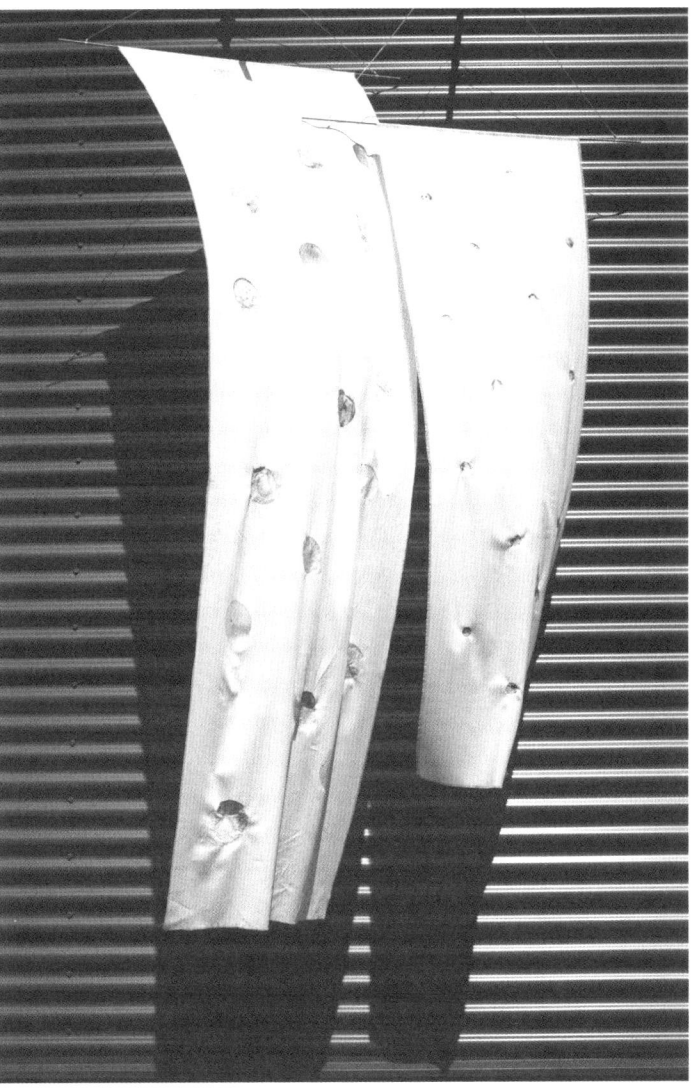

Bacteria Patterns heat pressed on textile © Tamara Hoogeweegen

2014_18

One Wash explored the idea of developing and using a bio based water soluble material to reduce the environmental impact of fast fashion. The garment prototypes from the *One Wash* collection disintegrate in the laundry after the wearer is done with them, thus leaving no textile waste after being discarded by the (fast fashion) consumer. The focus of *One Wash* is not so much a technological vision, as a questioning of on one of the major drivers of waste in fashion: The current mismatch between the product life cycle of a fashion product, and the short use phase of fast fashion. *One Wash* offers an alternative vision for the use of fashion in everyday life, where post-consumer waste is completely eliminated. The short movie in which *One Wash* is presented helps the audience envision how a product like this would be used and influence their life. Collaborators: Fashion Futures, by-wire.net, API-institute, TUe, Saxion and Aiming Better.

fashionfutures.org
youtu.be/NRjk8TYIwlY

One Wash wearing and washing process

One Wash

Anke Jongejan

One Wash Garment

2014-15_19

The *Bright Jacket* features tiny LEDs that can be turned on or off. The jacket features Imec and Holst Centre's flexible smart fabric interconnect technology, as well as miniaturized printed electronics. While the result is purely cosmetic, the project shows how technology can be integrated within a garment. This project and earlier projects that by-wire.net created for Holst Centre boosted the development of flexible integrated technologies for fashion.

by-wire.net/bright-jacket-for-holst-centre/

Bright Jacket

Holst Centre
by-wire.net

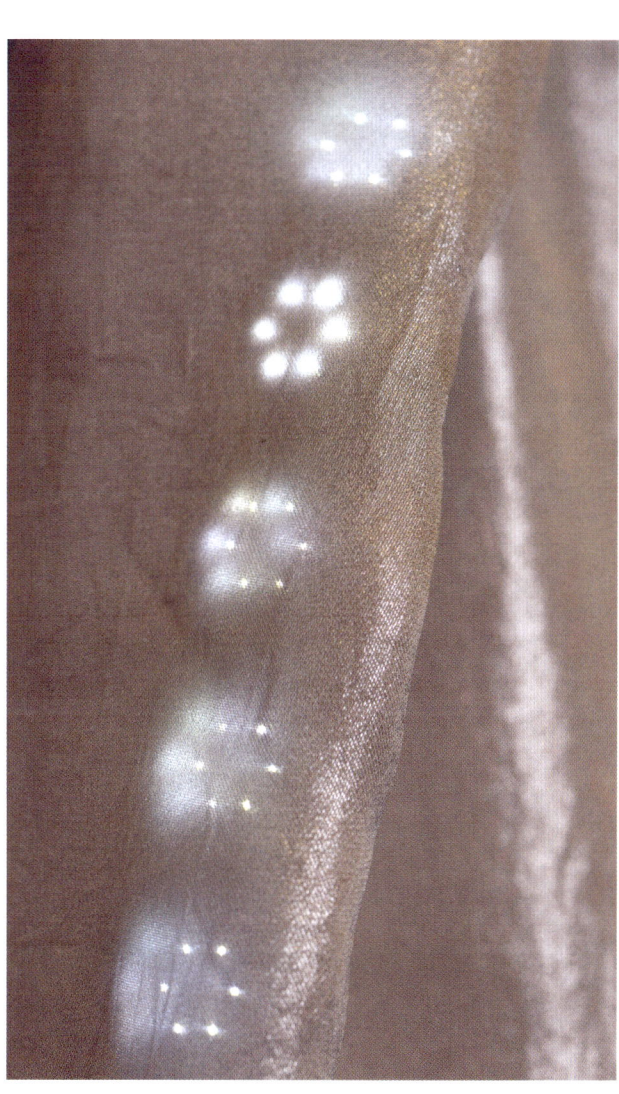

2015_20

Learning a physical activity, such as fencing, requires a combination of correct position of the body, feedback and repetition. To appeal to bodily skills to support the learning experience, *Flow* proposes a way of communication between wearable material and the body by means of directional cues delivered through inflatable pockets. The integrated air pockets and airways operate as material extensions of the actuators responsible for the inflation; allowing for exploring an aesthetic.

vimeo.com/133375840

The wearable *Flow*

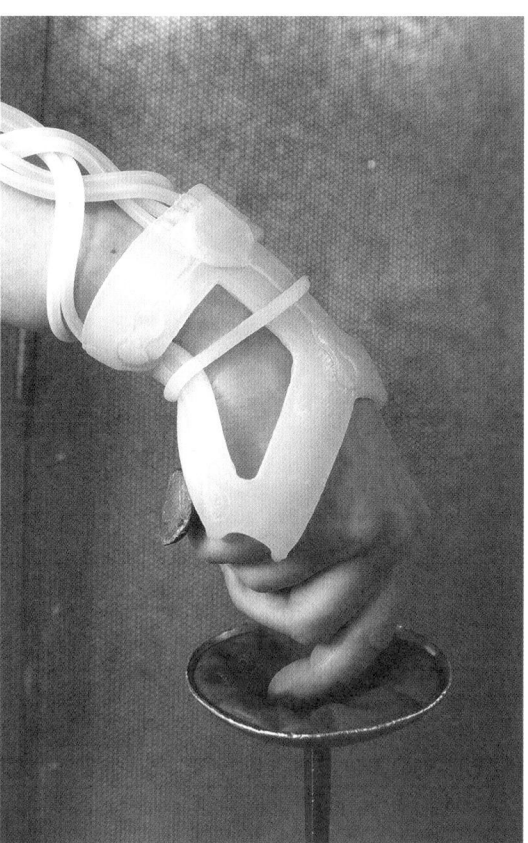

Flow

Bruna Goveia Da Rocha

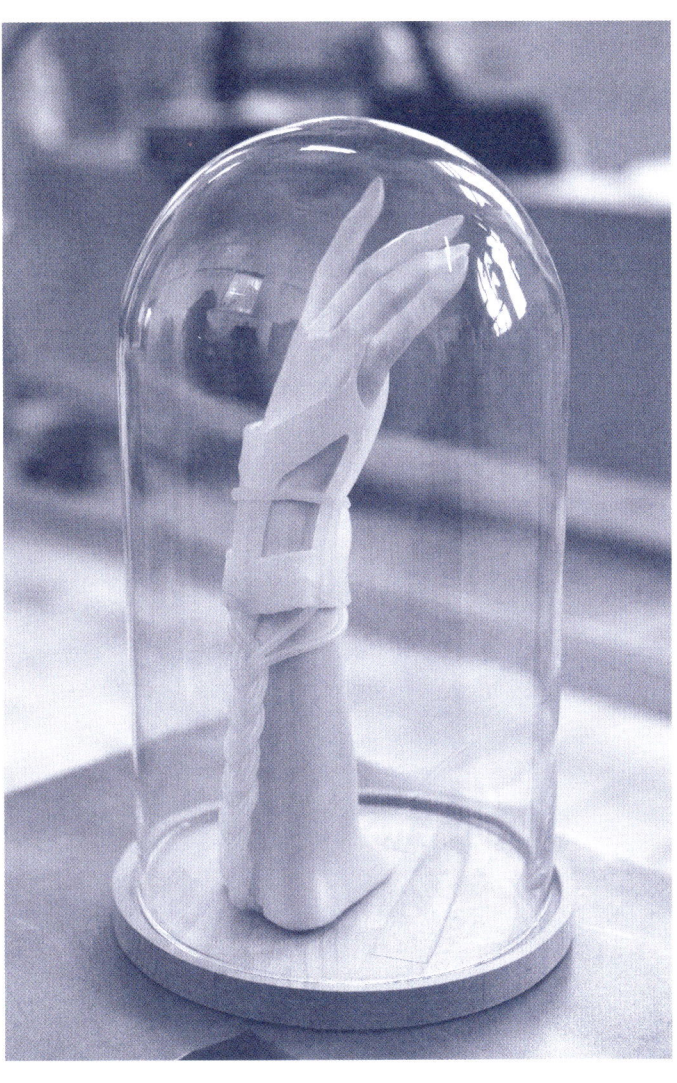

Flow presented during the Fashion? Future design for the present, Dutch Design Week 2018, photo by Armando Rodriguez Pérez

2015_21

Fashion on Brainwaves visualizes your brainwaves into morphing fashion. This morphing fashion is a second skin (of the brain) and a way to investigate the new aesthetics and beauty ideals (an extension of cosmetic surgery) applied to fashion and self-expression, which could lead to a new way of communicating.

In the *Fashion on Brainwaves* collection, eight basic brain frequencies are picked up through EEG technology and translated directly into either light, waveforms or shapes on the body. It can be seen as a direct symbol of a neural network. Building this was the first step to discover new ways to express our identity, to find a new way of communicating and adding extra functions to our daily garments through the combination of fashion, technology and science.

www.jasnarok.com/
youtu.be/8nYgOVH9VcU

Braight could be seen as a brain visualizer. The color of the garment will change on how you're feeling. Photos by Oona Smet

82

Fashion on
Brainwaves

Jasna Rok

Exstanding waves picks up 8 basic brain frequencies and translates this directly in 8 waveforms on the body distributed evenly over this shaped surface. Photos by Oona Smet

2015_22

Kimbow is an interactive dress which senses your posture and changes colour to amplify the message your body brings across. When standing in akimbo, with hands on hips and elbows projecting outwards, the thread structure of the dress changes colour in a way that creates an outward movement. This dynamic enhances the powerful appearance of the posture, attracting more attention and potentially increasing the wearer's confidence.

Kimbow is the result of an interdisciplinary collaboration between fashion technology designer Eef Lubbers and fashion designer and researcher Malou Beemer. With the support of Eindhoven University of Technology, TextielLab Tilburg and Aartsen Elektronica.

maloubeemer.com/project/kimbow/
vimeo.com/129483770

Unfolded garment to reveal the technology, photo by Bob Mans, model Janneke Jeurissen, muah Janine Van Helden

Kimbow

Eef Lubbers
Malou Beemer

Detail of the color changing front part, photo by Bob Mans, model Janneke Jeurissen, muah Janine Van Helden

2015_23

Crafting Wellbeing: Using Digital Technologies to Support Co-Designed Lace-Making Processes

chapter v

By Gail Kenning

Since 2014, I have visited the Netherlands many times and engaged in fruitful conversations with many of the pioneers of wearable tech. This has prompted me to think about the range of developments that have taken place, and to reflect on my own relationship to textiles and technology. My current work explores creative practices, participatory design and co-design approaches working with older people and people living with dementia. This interest stemmed from a series of projects that explored how older members of the community relate to craft activities. For many people, particularly women, their textile craft activities bring meaning to their life.[1] Craft-based textile activities such as knitting, crochet, and lacemaking provide challenges, physical and mental stimulation, creative outlets, and social interaction for many women. However, many people who have engaged in craft activities throughout their life and want to continue to do so, find that as they age they lack the dexterity and physical ability required. Not being able to make can impact their sense of self and wellbeing.

2015-18_30

Increasingly smart textile and innovative fashion tech projects, aim to support the wellbeing of people living with physical or mental health access needs or chronic conditions. Such projects have addressed autism in children [2016_31], and supported older people [2015-18_30] and people suffering pain [2010_09]. These projects operate outside of the mainstream fashion industry which so often focuses on idealised bodies and minds, and on the production of a 'normalised' concept of beauty. In contrast, the work of fashion-tech designers is often positioned as not engaging with beauty and aesthetics, and creating products that are perceived as alienating, stigmatizing, aesthetically impoverished and lacking in social wearability.[2] But opportunities for engaging with good design, beauty and aesthetics should be available to all. The 'Pattern as process' and 'Evolutionary lace' projects [2009-16_06] explored lacemaking and lace. This form of textile has been highly prized for its aesthetic beauty, and so the projects explored how this

2015-18_30 Cliff,
Mohamad Baharom

2013_06 Pattern as Process

form of engagement could be made accessible to all. The projects were premised on the idea that creative and engaging activities that require aesthetic judgement, can positively impact health and wellbeing.

Research across a range of disciplines, including arts and crafts, medical and health, and social studies, has begun to show the importance of 'everyday' creativity, such as craft-textiles, for wellbeing.[3,4,5] This form of 'everyday' creativity is available to all, including older members of the community. Maintaining the wellbeing and 'quality of life' of the older population, using non-pharmacological approaches, has individual, societal and economic implications. It is particularly needed at this time when the number of people over the age of 65 is globally set to increase threefold by 2050. In this way, the garment and textile industries can support wellbeing for older people by, firstly designing garments that engage with specific health needs and secondly, providing activities and 'things to do' that are associated with the aesthetics of fashion and textiles. My design and research projects began to explore how technology can be be used in relation to textiles to support the health and well-being of older people, particularly women involved in craft activities.

The research design projects attempted to disrupt thinking with regard to what constituted lace. My work began by exploring crochet lace making, a form of lace that had not been impacted by the industrial revolution, because the process could not easily be mechanised. Historically, lace had been thought of as a constructed textile formed from the conjoining of the material (thread) and the immaterial (space); it was a physical process producing a physical object to create and adorn clothing. By translating the patterns into a digital environment and engaging with the pattern forms at their systematic core, this research provided a deeper understanding of lace patterns and allowed for an exploration of whether a pattern's developmental path could be altered to create new emergent patterns [2013_06]. A similar process was used during the evolution of the Pied-de-poule [2013-17_14], a project by Loe Feijs and Marina Toeters. The 'Evolutionary lace' and 'Pattern as process' projects showed how crochet lace-making as an activity, could be revolutionised by using technologies that enabled patterns to be constructed

2013_06

digitally as code, and then physically produced using 3D printing.

The speculative lace patterns prompted questions as to why lace, when used in the fashion industry to adorn garments, was so often simply a refashioning of existing lace patterns, rather than an exploration of new patterns. Digital printing created the ability to print complex imagery to be used on fabrics, so what is the equivalence of this in constructed textiles? These projects suggested that it was in the translation of patterns into code combined with current 3D printing developments. This potentially opened up possibilities for technology to impact wearables both in the materiality of lace, and in shaping how lace is envisaged, shaped, valued and worn [2017_37], [2018_47], [2018_49]. The traditional value of lace was as an object and a highly prized accessory, the contemporary value of lace may be located in speculative aesthetics, in the process of *making*, or in the co-creation opportunities made available through the introduction of technology.

Using technology, a digitised form of pattern-making can be made available to older people who no longer have the dexterity to manipulate fine threads. Using a participatory approach a designer can provide support to enable someone to continue to do the activity that brought them joy, meaning, and a sense of community, and to enable them to engage in their craft activity and to continue to participate. Engaging people in participatory design and empathic co-design approaches provides opportunities to create wearable tech that is socially accepted and does not stigmatize, and for wearable tech developers to gain a greater understanding of aesthetics requirements and cultural needs.

2017_37

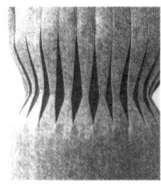

2018_49

2017_37 Perflex

2018_49 UNSEAM

Gail Kenning, *Senior Researcher at the University of Technology Sydney, Research fellow at the Ageing Futures Institute, University of New South Wales. Honorary Reader in Design for ageing and dementia at Cardiff Metropolitan University and the University of New South Wales.* Gail Kenning is an artist, designer and researcher exploring creativity, digital media, craft, expanded textiles andwellbeing. She is Research Fellow at Ageing Futures Institute, senior researcher at the Materializing Memories research program, University of Technology Sydney and researcher at Art and Design, University of New South Wales. Gail was awarded a PhD for her work exploring evolutionary patterns and code in relation to craft-based textile forms (www.gailkenning.com). She is very aware of the quality and impact textiles have on our everyday life and uses co-design and participatory approaches to engage with people often from a crafter and maker perspective.

1
Kenning, G. (2015). *"Fiddling with Threads": Craft-based Textile Activities and Positive Well-being.* Textile, 13(1), 50-65.

2
Dunne, L., Profita, H., & Zeagler, C. (2014). *Social aspects of wearability and interaction.* In Wearable Sensors (pp. 25-43). Academic Press.

3
Corkhill, B., Hemmings, J., Maddock, A., & Riley, J. (2014). *Knitting and Well-being.* Textile: The Journal of Cloth and Culture, 12(1), 34-57.

4
Csikszentmihalyi, M. (1996). *Creativity: Flow and the Psychology of Discovery and Invention.* 1st edn. New York: Harper-Collins Publishers.

5
Gauntlett, D. (2011). *Making Is Connecting: The Social Meaning of Creativity from DIY and Knitting to Youtube and Web2.0.* Cambridge: Polity Press.

6
Richards, R. E. (2007). *Everyday creativity and new views of human nature: Psychological, social, and spiritual perspectives.* American Psychological Association.

MycoTEX creates sustainable fabric from mycelium, the roots of mushrooms. This dress is the first proof-of-concept of a compostable dress. With a Body-Based modelling process developed together with Karin Vlug they create garments of this new textile that perfectly fit your body without the need to cut and sew.

neffa.nl/mycotex/
youtu.be/nVJv4bWnOCM

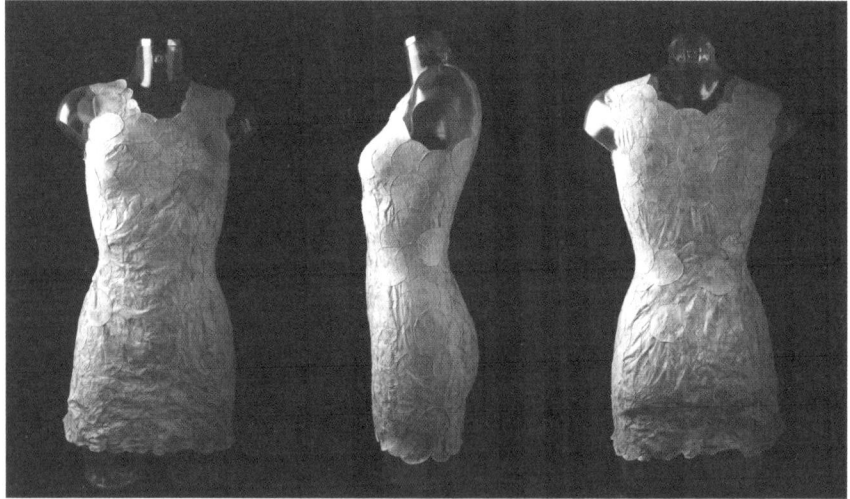

MycoTEX jacket developed with Karin Vlug © Jeroen Dietz

MycoTEX

Aniela Hoitink
NEFFA

2015_24

NazcAlpaca is a combination of alpaca yarn with wearable technology geared towards avoiding work related stress issues. The *NazcAlpaca* shirts monitor its wearer's body, and gives direct feedback via tiny vibration in the upper back. An app is used to adjust settings, start training and check your history. Two scarves can measure the air quality and temperature around the wearer. The *NazcAlpaca* project, developed with Bear Creek Mining S.A.C. in Peru, illustrates how data measured via integrated technology can actually be used on the body itself, rather than exist as a static base of knowledge used only for analysis.

by-wire.net/nazcalpaca-body-monitoring-alpaca-fashion-innovation/
youtu.be/NJr6n7fUba8

Dress back, photos by Iztok Klančar, model Kristel van Walen

NazcAlpaca

by-wire.net,
Marina Toeters
Martijn ten Bhömer

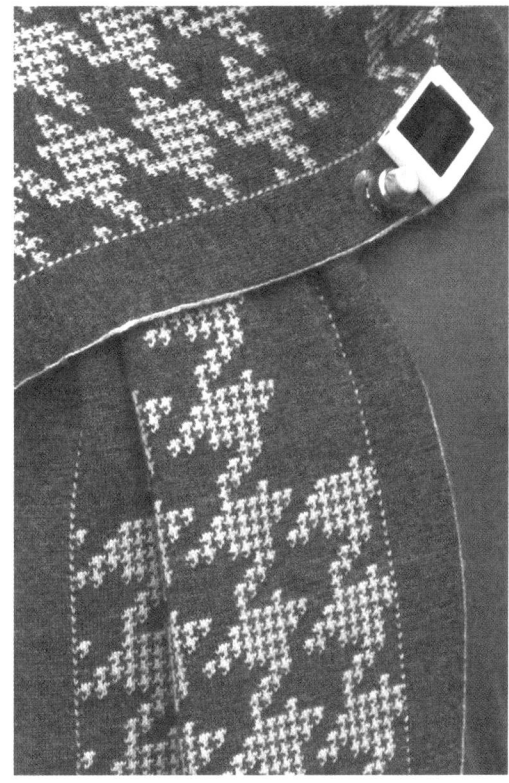

Detail NazcAplaca

Magnet connections and conductive yarn combined with alpaca yarn

2015_25

Phototrope is an illuminated running shirt designed to enhance a runner's ability to respond to changing light conditions. The design process started from the notion that light could not only improve safety but also contribute to a new form of aesthetic expression. It allows the wearer to remain visible and feel safe during a nighttime run, while the light also enables new playful interactions with other runners. With the support of by-wire.net *Phototrope* was user tested by a group of runners. This test revealed how the light would not only impact runner's physical performance but would, in some instances, also support them on a psychological level by enabling them to get into a flow.

Developed in collaboration with Philips Research, by-wire.net, Fred van Mook & the Lichtlopers and Aartsen Elektronica.

paulinevandongen.nl/
vimeo.com/127916473

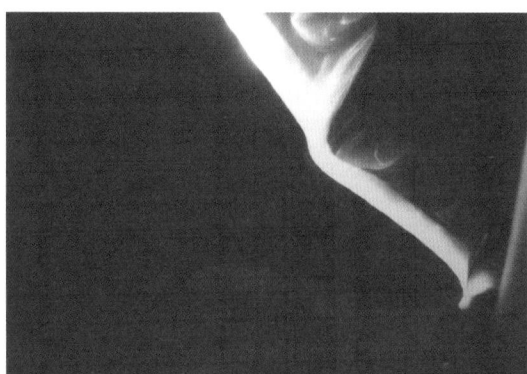

Photos by JR Hammond, Hammondimages

Phototrope

Pauline van Dongen

2015_26

The *Zishi posture sensing garment* monitors its wearer's posture, and uses the gathered information to improve therapy. The posture monitoring and correction technologies can support prevention and treatment of spinal pain or can help detect and avoid compensatory movements during the neurological rehabilitation of upper extremities, which can be very important to ensure the effectiveness of therapy.

by-wire.net/zishi-posture-sensing-garment-for-rehabilitation/

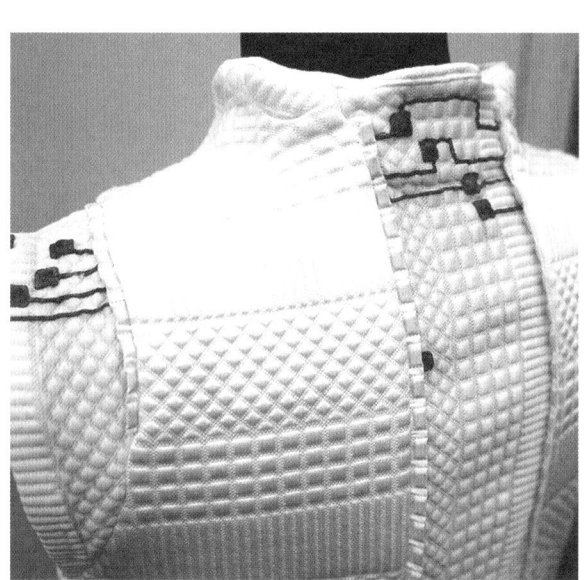

Back detail of the sensor connections, photo by Bart van Overbeeke

Zishi Posture
Sensing Garment

Qi Wang,
Eindhoven University
of Technology

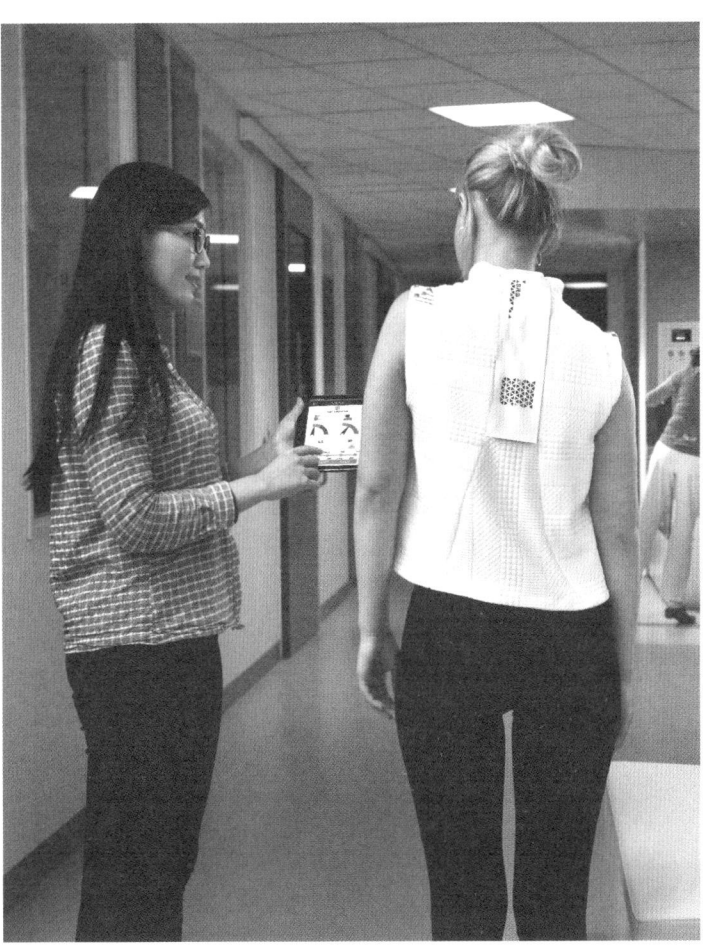

Photo by Bart van Overbeeke

2015_27

Fashion designer Karin Vlug researches the future of garment production by collaborating with partners and knowledge institutes. Contributing to a more sustainable fashion industry through material research and by redesigning the production process of a garment; enabling local, on-demand and digital production. Together with Laura, a researcher at AMFI, the basic concept of *Smart Fashion Production* was developed. To create a new fashion production system, beyond traditional methods such as pattern making and sewing. Designs are based upon personal 3D body scans for garments with a perfect fit, using 3D moulds for the production so there is no leftover material or extensive handwork.

karinvlug.com/smartfashionproduction/

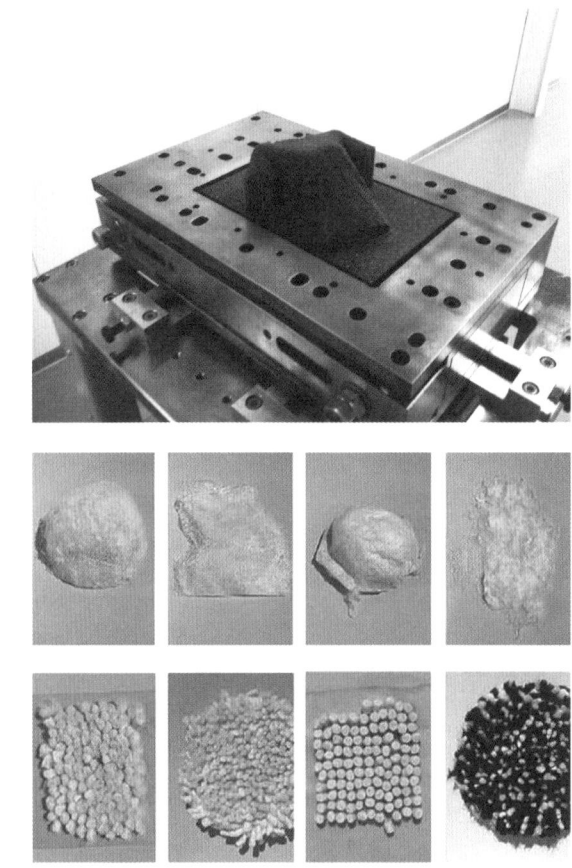

3D mould, Optimal Forming Solutions

Material explorations

2015_28

This Fits Me is a system that allows people to design unique and personalised fashion through 3D body scanning and generative algorithms. The system creates a virtual garment based on a 3D body scan of the customer, and the generative design of the garment can be customized. This way, the garment fits the body as well as the customer's identity. In the system, a generative line pattern is projected on the garment. By adjusting several variables in the generative algorithm, the customer can adjust the line pattern on the garment based on their personal preferences. The created line pattern will function as the seams in the garment, creating a new way of patterning garments. The pattern pieces are cut out of fabric with laser cutting technology.

leoniesuzanne.com
vimeo.com/107469973

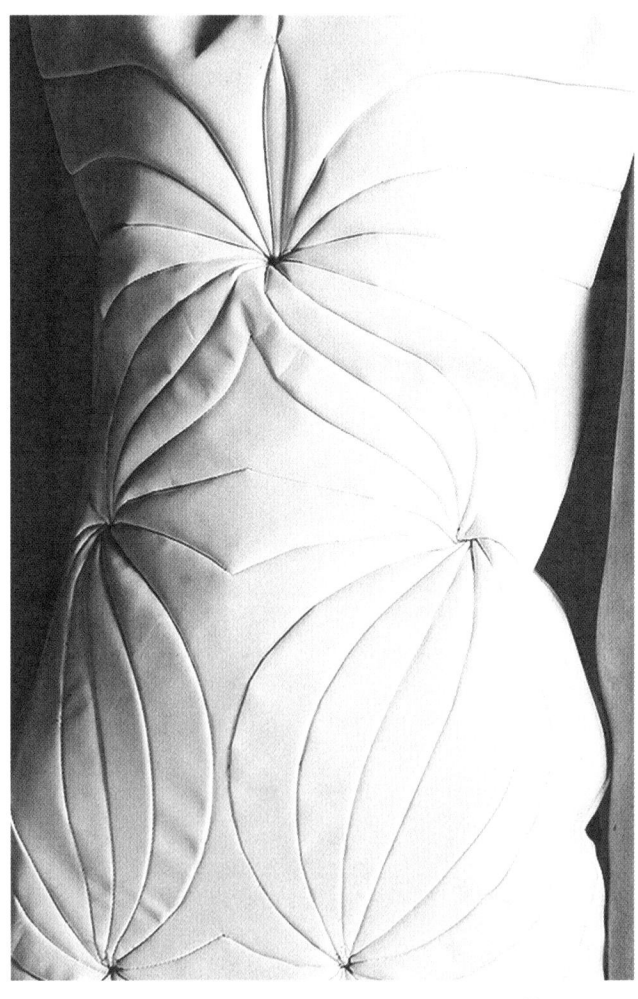

This Fits Me

Leonie Tenthof van Noorden
Eunbi Kim

Photos by Michel Zoeter. Hair and Makeup by Lisa Schuil. Model Amber @ Egomodels

2015_29

chapter vi

By Stephan Wensveen

The Wearable Senses Lab (WSlab) is a space for innovation, experimentation and education for close-to-the-body interactions that incorporate wearable computing or smart textiles, within the department of Industrial Design at the Eindhoven University of Technology (TU/e). The WSLab began in 2008, when we realised, through projects relating to electronics and textiles, there could be an interesting link between sports garments and electronics. Adidas AG was approached, recognized the potential in this, and provided financial support for the continuation of projects in this area. We purchased tools and materials, such as sewing machines and conductive fabrics, and thus the WSLab was born.

WSLab provides a platform and community for students, researchers and industry to create prototypes and explore the future together. This is achieved by combining research with education and by focusing on the development of skills in both research and practice.[1] We advocate a competency-centered and research-through-design approach. This approach can be seen as an iterative transaction between design and research in which skills, knowledge, and attitudes are generated through cycles of designing, building, and experimentally testing experiential prototypes in real-life settings.[2] At the WSlab, people from different disciplines work closely together. Interaction and fashion designers, people familiar with human physiology, psychologists, sociologists, mathematicians and engineers are all needed to create propositions that will be accepted by end users in the market. The WSlab has developed a strong network of regional, national and international industry partners, and receives support from both the field of textiles and the field of electronics.[3]

2011-15, 15 Vigour, TU/e
Wearable Senses Lab and many others

2011-15, 15 Spine Warming
Dress, TU/e Wearable Senses
Lab and many others

The role of fashion within interaction design according to the WSlab

The history of the Industrial Design department at TU/e is mired in technology and interaction design. Bringing fashion into the context of technology and interaction design meant we no longer had to remind students about the importance of the body; culture and aesthetics; attention to detail; choice of materials; or presentation. When fashion meets technology these aspects become self-evident and are inherent goals in the design project. Fashion allows for a similar kind of forward-thinking and advanced exploration offered in research for space or the military. It has a future forward role where usability or functionality are part of the design, but not the starting point. This provides room for material exploration and experimentation. Similarly to the meaning of concept cars for the automobile industry and designers, fashion provides this area of experimentation for interaction design and designers.

CRISP Smart Textile Services

2011-15_15

Each semester again, many projects and ideas were generated by our students, but they could not make a lasting impact yet. However, our reputation grew and we were invited to take part in the 4-year nationally funded Creative Industries Scientific Program [2011-15_15]. As a result, Smart Textile Services (STS) was set up. It was a collaborative project to create an 'inspirational test-bed' to explore platforms, methods, tools, and materials and work with multiple partners to produce increasing qualities of prototypes. This way the creative industries contributed with their expertise to explore the opportunities and challenges to design Smart Textile Product Service Systems.

2014_15

The methodological drivers for STS were the 'Growth Plan'[4] and our strong attitude towards prototyping and research-through-design processes. Within research-through-design the designer is the researcher and generates new knowledge in their design and realization processes of ideas, concepts and prototypes. However, STS extended beyond the traditional approach of one design-one designer/researcher-one context. Instead, the final portfolio was made up of eleven design projects [2014_15] from multiple designer-researchers targeting a diversity of user groups, such as the elderly [2013_15], parents and children, and dementia patients, their partners and therapists. Four of the eleven concepts involved multiple iterations [2013_15]. STS realised several industrial innovations, including innovative business models for technology partners; moving away from consultancy roles and introducing a technology platform for wearables; offering a renewed outlook on the innovative potential of the Dutch textile industry; and exploring the potential for cross-over collaboration between service providers, technology and textile partners. The project went beyond 'fuzzy front-end' prototyping and developed a growth plan for concepts from initial nursery, incubation through to adoption. It also introduced a new class of smart textile product service systems.[4]

The CRISP project provided different insights for each participating party. An important point that emerged was that the new technological opportunities of smart textiles needed to be understood in the context of their systemic nature; opportunities influence not only the product, the service and the user, but also everything around the user, including the production, maintenance and afterlife. Researchers were able to apply their research in the 'real world' of the partners. For example textile industry partners learned innovative approaches for product service systems, apart from what had been the traditional focus on metres of production. Service providers indicated how best to involve them in innovative research by, for example, approaching them systematically and not only providing a single piece of textile, or a single touch point in the product-service system.

2015_26

2015_26 Phototrope, Pauline van Dongen

The reputation of STS sparked new research projects, including NWO project 'Crafting Wearables' with results like Phototrope [2015_26], and Issho [2017_36], and a European Program ArchInTex with results like Phem [2019_50],[5] which is currently continued in a project on Ultra-Personalized Products and Services[6].

2019_50

2019_50 Phem, Angelia Mackey

Future aims: Research, education and society

Researchers, students, industry and consumers all tend to stay in their respective and convenient bubbles. They cannot, and should not tackle their challenges on their own. They should cross over and be confronted with multiple viewpoints in order to combine the idealistic and realistic, the short term and long term, the personal and the social.

1
Wensveen, S. A. G. (2018). Constructive design research. Eindhoven: Technische Universiteit Eindhoven.

2
Koskinen, I., Zimmerman, J., Binder, T., Redstrom, J., & Wensveen, S. (2011). *Design research through practice: From the lab, field, and showroom.* Elsevier.

3
Tomico, O., Wensveen, S., Kuusk, K., ten Bhömer, M., Ahn, R., Toeters, M., & Versteeg, M. (2014). Day in the lab: Wearable senses, department of industrial design, TU eindhoven. interactions, 21(4), 16-19.

4
Wensveen, S. A. G., Tomico, O., ten Bhomer, M., & Kuusk, K. (2015). Growth plan for an inspirational test-bed of smart textile services. In *10th ACM Conference on Designing Interactive Systems (DIS 2014), June 21-25, 2014, Vancouver, Canada.* Association for Computing Machinery, Inc.

5
Mackey, A., Wakkary, R., Wensveen, S., Tomico, O., & Hengeveld, B (2017, March). Day-to-day speculation: Designing and wearing dynamic fabric. In *Proceedings of the Conference on Research Through Design* (pp. 439-454).

6
Ten Bhömer, M., Tomico, O., & Wensveen, S. (2016). Designing ultra-personalised embodied smart textile services for well-being. In *Advances in Smart Medical Textiles* (pp. 155-175). Woodhead Publishing.

For a university, the future of societal relevant education is in challenge-based learning, where students learn through authentic and open-ended projects. In these projects they collaborate with researchers, the industry and members of society to tackle the societal challenges and explore the future. These projects can inspire ambitious and idealistic students to initiate real change, [2014_18], [2015_21], [2015_23], [2015_29], [2016_34], [2017_37], [2018_42], [2018_44].

A platform like the WSlab allows for two perspectives on change, the shorter viewpoint of the society we have now and, and further up, when developments continue and change the future. For these viewpoints we need to integrate technological developments and research into a societal context, and study the opportunities and consequences together. We need to be forward looking, and research the future by realizing it at the same time—by doing, analysing and framing knowledge as it is acquired. We at the Wearable Senses Lab are happy to collaborate with you. Let's explore the future together!

Stephan Wensveen, *Director of Education, Professor Constructive Design Research, Eindhoven University of Technology.* Stephan Wensveen is Professor Constructive Design Research in Smart Products, Services and Systems and Program. Director for the Bachelor's degree and graduate programs of Industrial Design, Eindhoven University of Technology (TU/e). His interest is in using the power of research through design to foster collaboration between research, education and societal innovation. Stephan is the initiator of the Wearable Senses Lab (2008) and consortium builder of the CRISP Smart Textile Services program [2011-15_15]. Stephan generously supported the DDW18 expo and motivated to develop this book.

The elderly, people with a disability or even people wearing tight-fitting clothing can experience difficulties when opening and closing zippers, because they are physically unable to do so, or because the zip is located on the back of the garment. *Cliff* helps people to open and close their zippers and is here applied on a sublimation print dress. *Cliff* is a design approach towards fashionable automated assistive technology. The design can also be used for other industrial applications such as tents.

by-wire.net/cliff/

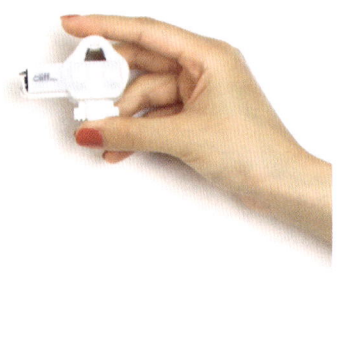

The fifth iteration prototype worn by Joke Loozeman on the Yellow Sublimation Zipper Dress, photos by Maxime Dassen

106

Cliff:
An Automatized
Zipper

Mohamad Zairi Baharom

2015-18_30

Agent Unicorn is an accessory shaped like a horn on a unicorn, for children with ADHD. The horn-shaped headpiece measures brain activity through the P300 wave. A little camera inside the unicorn horn gives the children extra eyes. The headpiece measures their attention level and focus. Children with ADHD often have problems with both. The headpiece is designed to help find out what might trigger them and gives a better understanding of their individual distractions. It is also a very playful device that gives these children a little bit more of a playful aspect in the medical environment.

anoukwipprecht.nl/
vimeo.com/174628551

Photo by Marije Dijkema

Agent Unicorn

Anouk Wipprecht

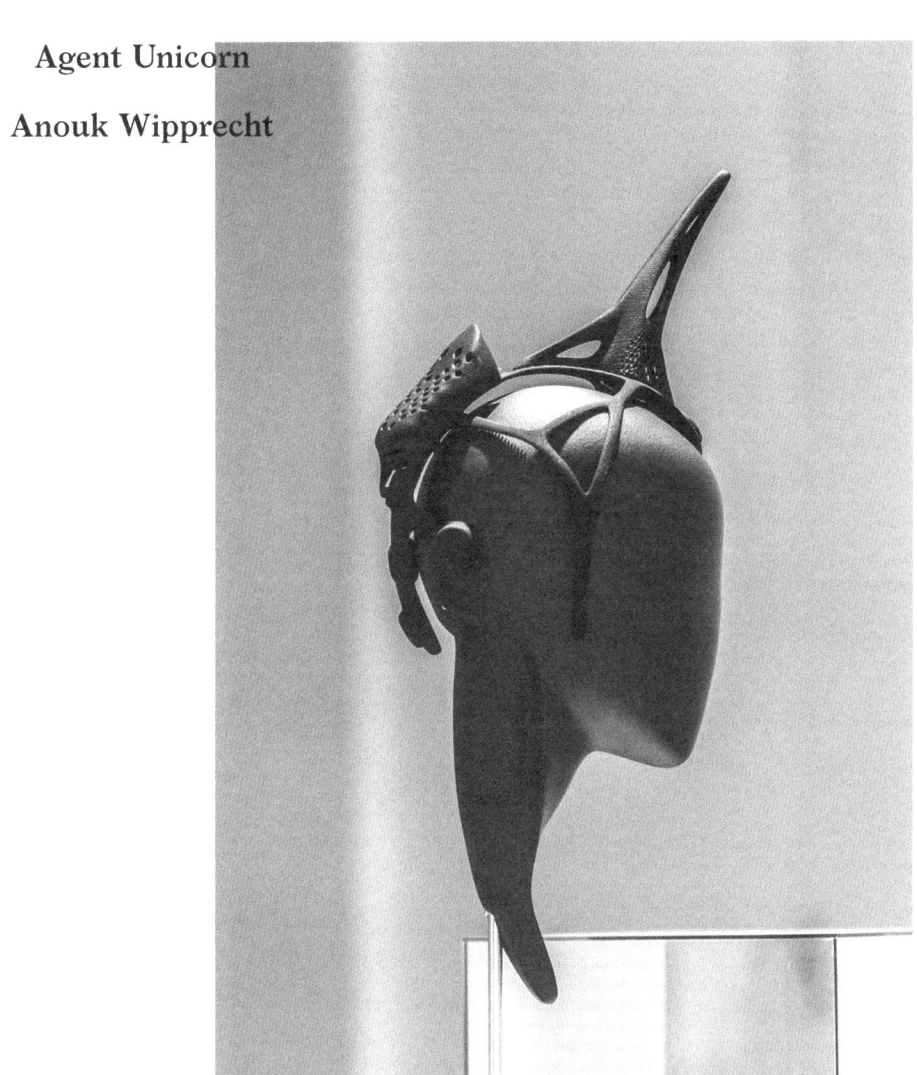

The Dutch MedTech start-up company *Bambi Medical* develops its first product a wireless neonatal vital sign monitoring system—the Bambi Belt. The Bambi Belt effectively measures vital signs (ECG, respiration) of premature babies in a skin friendly and wireless way. The Bambi Belt aims to help reduce the pain and stress that is present in the current way of monitoring of premature babies. It will also facilitate easier handling by nurses and parents for optimized Kangaroo Mother Care, the skin-to-skin contact between parents and baby, which according to the World Health Organization has proven to be essential for parent-child bonding while Improving long-term clinical outcomes. Bambi is based on the PhD research of Sibrecht Bouwstra (2013) TU/e.

bambi-medical.com/
youtu.be/aSpJOzTM744

Bambi Belt

Bambi Medical
Sibrecht Bouwstra

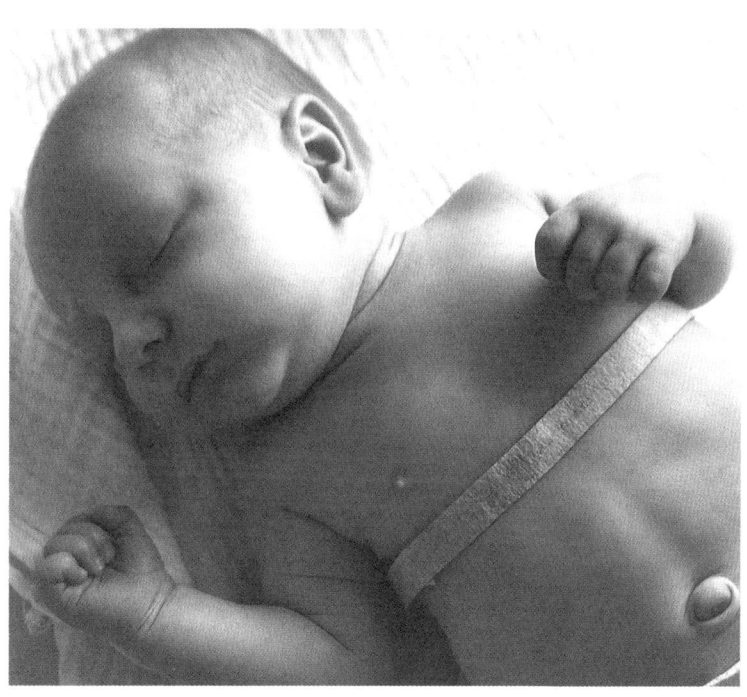

The Bambi Belt in use

In collaboration with the ILJA team, by-wire.net developed integrated light systems for the SS16 couture collection *ASSIMILA*. These light systems were integrated in five couture looks, which were shown at Paris Fashion Week. As its name implies, the *ASSIMILA* collection focuses on the concept of 'assimilation'. The collection was inspired by the skin, where the change of colour is poetically expressed by the integrated light systems.

by-wire.net/ilja-assimila-ss16-fashion-technology/
youtu.be/KGdxrNUOHCA

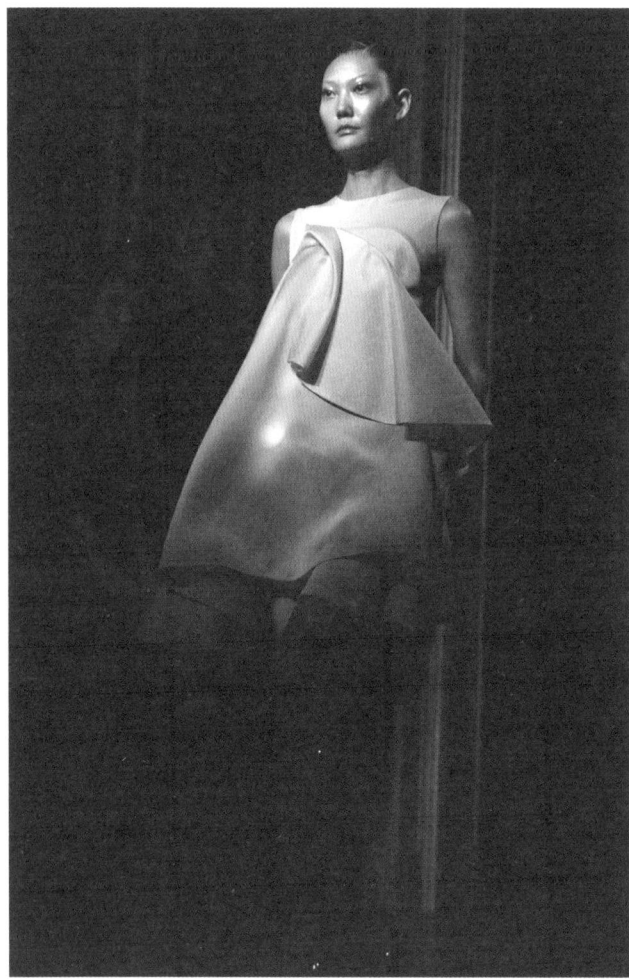

Photos by Elise Toide (left) and Peter Stigter

ILJA ASSIMILA SS16:
Couture with
Light Systems

Ilja Visser
by-wire.net

2016_33

Spellbound is an Estonian-based brand which describes its audience as 'conscious magic-lovers'. The products are designed to shape and influence attitudes on sustainable consumption of fashion products. The items are developed and produced in Estonia by using high-quality locally sourced leftover materials and ethical principles for production. Additionally, *Spellbound* explores new sustainable directions in textile and garment design, while featuring interesting twists by the implementation of technology.

spellbound.ee/

Spellbound

Kristi Kuusk

Photos by Worth project. illustration by Kerstin Zabransky

2016_34

SaXcell is an innovative cellulose fibre produced from cotton waste
SaXcell, an abbreviation of Saxion cellulose, is a regenerated virgin
textile fibre made from chemical recycled domestic cotton waste.
The process to transfer domestic cotton waste into *SaXcell* fibre is
a crucial step in the circular textile chain.

saxcell.nl
youtu.be/-27Moxcwm-Q

SaXcell fibres

SaXcell

Saxion

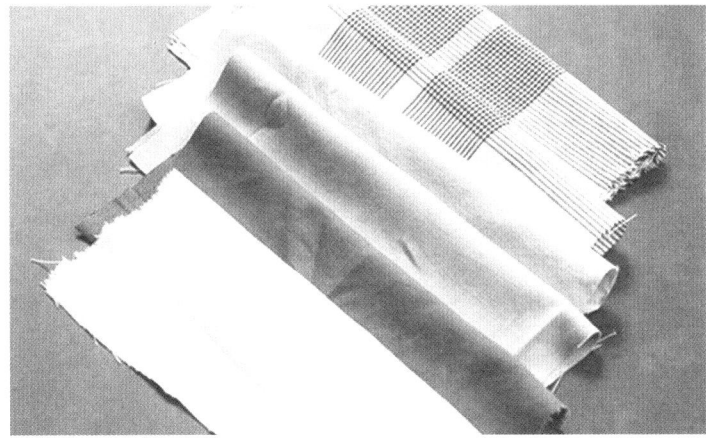

Issho means 'togetherness', 'same' and 'shared' in Japanese. *Issho* started with the newfound possibility of weaving conductive yarns into denim fabric to obtain a level of intelligence enhancing the body in relation to its daily surroundings. Golden conductive yarns that are incorporated in the weft of the fabric create touch sensitive zones. The intelligent denim jacket records social interactions (physical encounters and activity on smartphones) and is also able to give feedback to the wearer using small vibration motors. Through bodily sensations—the feeling of a gentle stroking on the upper back—this jacket encourages the wearer to be present in the moment.

paulinevandongen.nl
vimeo.com/207468324

Issho

Pauline van Dongen

Photos by Sharon Jane Dompig, muah Angelique Stapelbroek

2017_36

From Design Fiction to Design Future: Bringing Sustainable Ideas and Innovations Closer to Reality

chapter vii

By Anke Jongejan

Innovation in fashion is a complex affair. The fashion industry has proven itself to be ill-equipped when it comes to initiating the changes necessary for a sustainable future. This is for three reasons. The first is the speed of product turnover. The second reason is lack of focus. Fashion companies usually have no research and development departments to ensure broad forward thinking. As a result, most companies make no connection between emerging innovation and day-to-day decisions. Their focus is on the forecasting of styles and colours and maintaining sales. This means that change in fashion is limited and often superficial; one style simply replaces the next.

The final major reason for the fashion industry not being well-equipped to initiate change is due to the end user, the wearer. Fashion largely has a cultural function: through it we experience our identity, and communicate it.[1] As users we experience a strong connection between what we wear, who we perceive ourselves to be, and how we think others will see us. In this context, changing styles is scary enough, and so changing paradigms can be overwhelming.[2] This means that fashion companies are dealing with audiences that are not necessarily interested in wearing new products on their body. For most people, fashion and clothing simply are an area of conformity rather than one of innovation or even revolution.[1]

Speculative design as provocation

So how can we create this paradigm shift? Change in fashion cannot be made without imagining a better way, or a better future, and for this purpose speculative design is a powerful tool. Speculative design does not try to predict the future, but imagines and visualizes possible future scenarios. It uses design to "open up all sorts of possibilities that can be discussed, debated and used to collectively define a preferable future [...]".[3] Conversely, today's fashion industry insiders and experts *do* try to predict the future, trying to understand the future wants

2014-15_19, One Wash,
Anke Jongejan

and wishes of an audience needing to be enticed into purchasing. While speculative design engages with current issues in society, it can *choose* the future it wants to aim for, and build on that. As Dunne and Raby say, "We are not talking about a space for experimenting with how things are now, making them better or different, but about other possibilities altogether".[3]

This kind of design may revolve around a futuristic proto-type, but it relies heavily on scenario building and story-telling, which is why the genre is also often described as Design Fiction. This quality gives speculative design the power to aid and guide the fashion industry in the trans-formation it will need to go through, while also creating a vision in which the wearer of fashion can feel, and become accustomed to, the experience of a new identity. Speculative design doesn't necessarily offer answers or alternatives. Its main focus is on prompting the viewer to question today's reality through 'unreality'. The designed visions of the future are 'provocations', meant to change the way we think.[4] Speculative design is often based on a kind of idealism to, "increase the probability of more desirable futures happening".[3] However, currently it seems to have moved from proposing 'far future concepts' to a focus on 'validation'. More and more designers take responsibility for the applicability of their ideas with what would realistically be possible to achieve in the next five to ten years.[4]

2014-15_19

Design Fiction in fashion

This idealism and a focus on validation are key drivers in the 'functional fictions' that we see in Dutch speculative fashion design. These future explorations use design and science to explore how technological advances could be adopted.[3]

    The One Wash [2014-15_19] project, which I worked on in collaboration with university researchers and industry partners, used scenario building and fictional prototypes to introduce lifecycle-thinking in design to both the fashion industry and wearers. Karin Vlug's Smart Fashion Produc-tion [2015_28]; the Posterus Textilus project [2014_18] by Tamara Hoogeweegen and Contre Choc and by-wire.net's Waste Conscious Scarf [2012_12], all use design to explore alternative scenarios based on technological (near-)possibilities, either for an industry audience or to directly impact consumer behaviour.

The Res Materia [2017_38] project by Sanne Karssenberg's does not seek validation but uses pure 'unreality' to explore the relationship of fashion with the 'new'. This use of speculation as artistic research shows the important role of universities and art schools as a breeding ground for new possibilities through 'unreality' and wild exploration. A great commercial example of speculative design is Studio Roosegaarde's proposal for Lacoste's Polo of the Future [2012_11]. The short movie demonstrates the potential for design fiction to experientially engage wearers of fashion with a possible future—fundamental for having an impact and remove barriers for change.

Functional fictions have an important role to play in bridging the gap between possible innovation and its use and implementation in the fashion industry. If used skillfully they have the power to resolve the three barriers to innovation defined earlier, and propel the fashion industry towards the desired future. Important future directions for this kind of design are in placing a strong emphasis on understandability and accessibility of the scenarios for both the wearer and an industry audience. Science and academia are crucial for generating ideas and building innovation, both in unreality and applicability, but for real impact on the fashion industry the accessibility demonstrated by Roosegaarde for Lacoste is the key to the future.

2017_38

2012_11

2017_38, Res Materia, Sanne Karssenberg

2012_11, Polo of the Future, m.nster. and Studio Roosegaarde for Lacoste

Anke Jongejan, *Director of the Fashion Department, University of the Arts Utrecht (HKU).*

Anke Jongejan is the director of the Fashion Department of University of the Arts Utrecht in the Netherlands. She also owns the company Fashion Futures. Fashion Futures aims to make tangible and real what could be via Speculative Design and the Design Fiction approach. Make abstract visions of the future tangible and real (www.fashion-futures.org) [2014-15_19]. On her LinkedIn page, she states how she finds 'fashion [...] the most fascinating field there is, with a lot of room for change (if you're up for a challenge).' Anke educates young fashion professionals toward being discerning actors in the field.

1
Barnard, M. (1996). *Fashion as Communication.* New York: Routledge.

2
Woodward, S. (2009). 'The Myth of Streetstyle' in *Fashion Theory.* (vol 13, iss 1). pp. 83-102.

3
Dunne, A and Raby, F. (2013) *Speculative Everything; design, fiction and social dreaming.* Cambridge: MIT press.

4
Dijksterhuis, E. (2015). 'De Grote omwenteling' in *Dude, Dutch Designers Magazine.* (ed. 4). pp. 38-42.

Commercial 3D-printed wearable products are already on the market today. Produced mostly by large companies, these products provide consumers with only limited options: 'One size fits most' rather than 'one size fits all'. *PerFlex* aims to take away this limitation by creating items that are 'unique size fits you'. This is done by combining a parametric pattern made by designers with the input of body data to generate a personalized 3D-product. Possible products range from shoes to underwear.

perflex.design

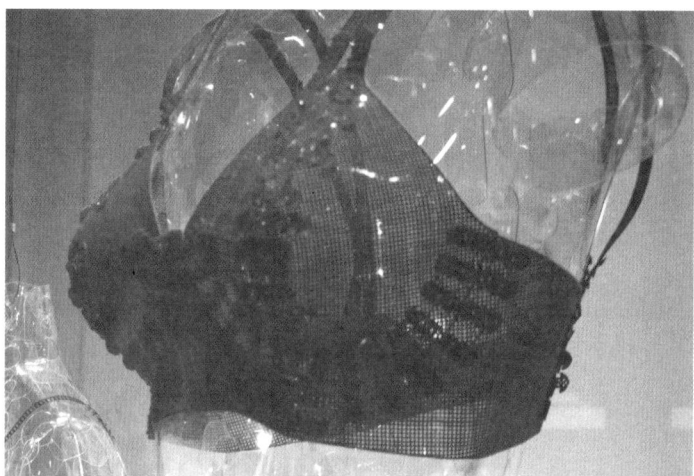

Perflex presented during Fashion? Future design for the present, Dutch Design Week 2018, photo by Armar do Rodríguez Pérez

3D printed underwear

PerFlex

Brigitte Kock
Bart Pruijmboom
Niek van Sleeuwen

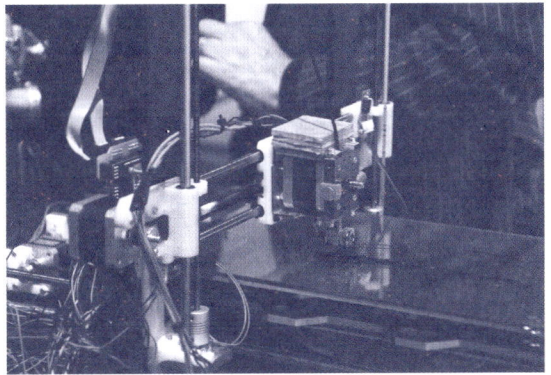

3D Printer photos by Sho Suzuki

3D printed bra photo by Sho Suzuki

2017_37

The installation *Res Materia* proposes a direct and personalized form of re-production of garments. The project gives a future perspective on a sustainable and embodied re-creation of clothes. Both the product and the re-creation process involve the wearer deeply. The project consists of the following protocol: a person brings in a garment with a history, a piece one would not throw away but also no longer wants to wear. The garment goes through a shredder, followed by the wearer stepping into a cabin that produces a whirlwind with the same remnant fabric fibers.

The storm of fibers creates a new and unique layer of textile on the wearer. Like this, garments become agents for change and transformation, instead of representing the new.

sannekarssenberg.nl
youtu.be/0uegtok3-i8

Res Materia Exhibition © Sander van Wettem

Res Materia

Sanne Karssenberg

Shredded Garments, Photography: G-Star

2017_38

The *Solar Trenchcoat* by Lithe Lab was designed as a part of the Aimey seated clothing collection. The trenchcoat can charge the wearer's devices and the batteries for the sensors and actuators in the other garments of the collection. When the wearer is outside in the sun, or even under bright office lights, they can put the solar panel on the backrest of the chair. The Aimey collection was designed for people who spend most of their time in a sitting position. The garments can sense if the wearer is sitting with the right posture, can massage the areas where pressure sores can form and heat up if the body can't regulate the body temperature.

Lithe Lab designs ultra personalised medical and supportive products. Functional and aesthetic possibilities are researched in collaboration with the wearer of the products.

lithelab.com

Back detail

Solar Trenchcoat

Lithe Lab
Daisy van Loenhout

Photos by Bianca Gorini

2017_39

The Philips *Smart Sleep* device promises its wearer that they feel wake better-rested and more refreshed, without changing how long they sleep. It does so by by tracking sleep patterns with the *SmartSleep* headband, in combination with sounds and an app that actively improves upon these sleep patterns. User feedback is reporting 80% improvement in their sleep. The very first prototypes were developed by by-wire.net.

usa.philips.com/c-e/smartsleep.html
youtu.be/d9E5O-AAd2M

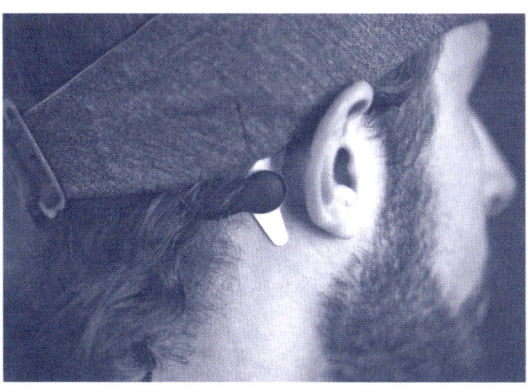

Smart Sleep, photos by Philips

A collection of swatches of interactive textiles and electronics that can be used for experimentation and education. Independent pieces of fabric with energy source and switch, ready for mounting on your garments to test the effects and functionality. Additionally, *Beam* is working on combining fabric and light. Using microcontrollers and sensors this combination becomes a symphony, illustrating both big and imperceptible feelings of being individual in a community.

flickr.com/photos/contrechoc/
youtu.be/Jom178hT7BY

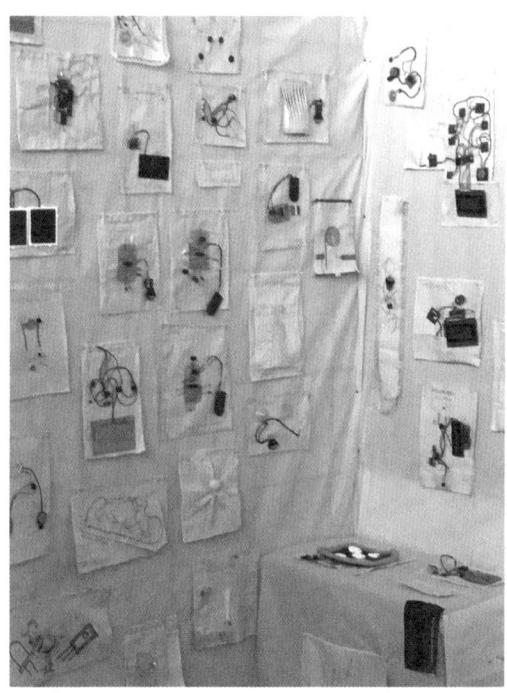

Portable Sensorlab presented during *Fashion? Future design for the present*, Dutch Design Week 2018, photo by Armando Rodriguez Pérez

Portable Sensorlab &
Weihnachtsbaum

Beam Contre Choc

Weihnachtsbaum presented during Fashion? Future design for the present,
Dutch Design Week 2018, photo by Armando Rodriguez Pérez

2018_41

LABELEDBY. is a research and technological development studio, based in the Netherlands, Eindhoven. *LABELEDBY.* offers design expertise, collaborative creative services and exclusive products, and is always on the look-out to explore and exploit the possibilities of new techniques as 3D printing and laser cutting. They believe it is the duty of designers to inspire, question the status quo, and bring history and future to the present while sharing this with society. *LABELEDBY.* created 3D printed buttons, buttonholes and finishing for Karin Vlug, to make garments that can be constructed by the consumer.

labeledby.com/

Fabienne and Jessica at work, photo by Jeroen Cox

LABELEDBY.

Fabienne van der Weiden
Jessica Joosse

2018_42

chapter viii

2017_36, Issho,
Pauline van Dongen

By Pauline van Dongen and Oscar Tomico

Today's society places great emphasis on technological development and innovation. And while we are indeed dependent on new technologies to overcome some of the global challenges we are facing, we sometimes risk putting technology on a pedestal. It can lead people, including designers and engineers, to think of new technologies as something more unique than the existing technologies. This technocratic approach arises when technology is treated differently from other materials. The prevailing understanding of technology is not that of a material, but of a tool or instrument. Especially in the case of wearable technologies we find this to be problematic. From this viewpoint, technology delivers added functionality and the rest of the design, a garment in this case, must serve this purpose. Arguably however, in everyday contexts, garments and accessories are experienced as material artefacts as much if not more than tools. Of course, technology enables new kind of interactions and brings along different material properties. But, it is important not to forget that all materials are in some way active, or to some extent responsive. For example, wool has thermo-regulating properties. By focusing on functionality there is a risk of installing a degree of determinism in the design process. Whereas, how people interpret and appropriate a piece of wearable technology can only be anticipated to a certain extent. Every wearer will experience the garment in their own way, depending heavily on the context of their experience.

2014_16, Solar Shirt,
Pauline van Dongen

The challenge for designers is to find ways to deal with this degree of uncertainty and to create garments that allow for a certain fluidity. We would like to encourage you, the reader, to consider not only what the garments presented in this book *do* functionally, but how wearing a particular garment may influence your felt-experience and perception of the world or how it may change your actions.

In the following paragraphs we would like to offer a starting point coming from our own personal research practice. CRISP Smart Textile Services and Crafting Wearables were two large Dutch research programs focused on a

2017_36

body-centred approach to wearable technology were we participated. As design researchers involved in these programs, we drew on phenomenology to inform our embodied practices with technological materials.[1, 2] The experiential body was central to projects such as Vibe-ing [2013_15], a garment that invites the wearer to feel, move, and heal through vibration therapy, Phototrope [2015_26], an illuminated running shirt that allows new interactions for group running, and Issho [2017_36], a jacket that can raise body awareness through the experience of a gentle stroke on the wearer's back.

Technology is a material

The main insight gained through developing these garments, and many others, was that using a hands-on approach turned technology into something that can be touched and transformed, just like any other material that designers typically work with. This meant that the process did not start with solving a technical problem, and it does not hold on to predetermined ideas about the technology. It showed that when designers look at a specific technology through a material lens, they do not think of it in a generalized and disembodied way.

2014_16

A solar cell for example, cannot be understood through its energy harvesting capabilities alone. Every type of cell has its own material properties, such as size, colour, and flexibility, which distinguishes it from others [2012_13]. The materiality of the cell is what gave shape to the garment. A thin-film solar cell [2014_16] and a crystalline solar cell [2017_39] offer the same functionality, but they are materially different. While solar cells may represent 'energy harvesting' and 'sustainability', their role in the design process was not confined to this functional representation.

Like with any material that is cut, moulded, assembled or draped, a designer can shape the technology by working with its physical properties and behaviour. A material may suggest a function, but the design is not limited to it. This way of thinking opens a design space in which solar cells and printed circuitry [2018_45] can become a graphic print on a T-shirt. Moreover, the experience of wearing the garment and how someone engages with it, goes beyond the generation of energy alone. Focusing on the subjective experiences of wearable technology thus revealed that technology is a material matter, more than a matter of function.

Mediators for social acceptance

When approaching technology as a material, designers and engineers can develop a more explicit and embodied understanding of technology. They no longer need to overcome the dichotomy between aesthetics and functionality. [3] One prominent way that is being advocated is that rather than looking at technologies as something opposed to humans, it is possible to consider technologies as *mediators* between human beings and their world. [4] Philosophy of technology, and in particular postphenomenology, provides a theoretical framework that design practitioners can borrow from. As mediators, garments shape subjective experiences that are always connected to the wearer's context. [5, 6] Runners wearing *Phototrope* during a group run will be able to communicate to each other through the light of their garments that responds to their actions. Each runner becomes engaged with the group through constantly shifting human-garment relations. The *material aesthetics* [6] of the garment contributes to the social dynamic. When we challenge ourselves to explore the sensorial and relational side of wearable technologies, new design spaces will open which may facilitate the appreciation and appropriation of wearable in the everyday.

Pauline van Dongen, *fashion designer and researcher specialized in wearable technology.* Her design studio is dedicated to collaborative, cross-disciplinary work that enables innovation and experimentation to take form in a wearable and desirable product. There is a strong focus on materiality in her work. She questions how new materials, like electronic components, conductive yarns and software, relate to existing garment technologies such as fabrics, textile coatings and zippers. Currently she is finishing her PhD at Eindhoven University of Technology within the Crafting Wearables project (2013-2018) about the practice of designing wearable technologies [2015_26]. Ultimately, with the aim to advance the practice of designing wearable technologies and the role of wearable technologies in the everyday context of fashion.

Oscar Tomico, *Head of the Industrial Design Engineering bachelor program, Elisava, Barcelona, Spain. Assistant Professor, Eindhoven University of Technology. Project leader Dutch Creative Industry Scientific Program (CRISP) – Smart Textile Services* [2011-15_15]. Oscar is working on soft interactions. His current projects focus on the textile industry and involve stakeholders during the design process to create ultra-personalized smart textile services in the form of soft wearables. Oscar shares his knowledge via lectures and workshops all over the world. Oscar Tomico guided Pauline's PhD from his position at the Eindhoven University of Technology, Wearable Senses Lab.

1
Tomico, Oscar, and Danielle Wilde. 2016. "Soft, Embodied, Situated & Connected: Enriching Interactions with Soft Wearables." mUX: The Journal of Mobile User Experience 5 (1). https://doi.org/10.1186/s13678-016-0006-z.

2
Smelik, Anneke, Lianne Toussaint, and Pauline Van Dongen. 2016. "Solar Fashion: An Embodied Approach to Wearable Technology." International Journal of Fashion Studies 3 (2): 287–303.

3
Joseph, Frances, Miranda Smitheram, Donna Cleveland, Caroline Stephen and Hollee Fisher, 'Digital Materiality, Embodied Practices and Fashionable Interactions in the Design of Soft Wearable Technologies', International Journal of Design, 11 (2017),

4
Ihde, Don, *Technology and the Lifeworld: From Garden to Earth*, Bloomington, IN: Indiana University Press, 1990

5
Van Dongen, Pauline. Forthcoming, 2019. *A Designer's Material Aesthetics Reflection on Fashion and Technology*. PhD Thesis, University of Technology, The Netherlands. ArtEZ Press.

6
Verbeek, Peter-Paul. 2005. *What Things Do: Philosophical Reflections on Technology, Agency, and Design*. Penn State Press.

This biodesign research project investigates a natural way of dyeing textiles by using bacteria that produce pigments. *Living Colour* is a sustainable alternative to artificial textile dyes and can be used for both natural and synthetic fibres. The process of "bio-dyeing" with bacteria produces a biodegradable dye with very little run-off, little water use and low dyeing temperatures, without the use of toxic chemicals, textile treatments or fixing agents. The bacteria can be cultivated on vegan nutrition from (agricultural) waste streams. Leftover pigment could be reused for products that require less saturated pigments.

livingcolour.eu/

Living Colour presented during the Fashion? Future design for the present, Dutch Design Week 2018, photos by Laura Luchtman

Living Colour

Laura Luchtman
Ilfa Siebenhaar

2018_43

Clothing sizes have been notoriously unreliable in the past fifty years, creating large quantities of returns and waste due to ill-fitting clothing. Additionally, some consumers put value judgements on specific sizes and whether their body adheres to these sizes or not. *The Sizeless Store* wants to present a reality where clothing sizes are non-existent. *TSS* uses mobile 3D scanning and new technologies to guide the consumer to the best fitting garment without the need of size codes.

biancagorini.com/the-sizeless-store/
youtu.be/586hFmBVuRM

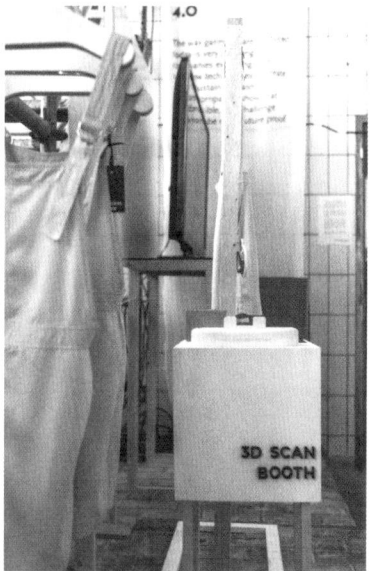

The Sizeless Store

Bianca Gorini

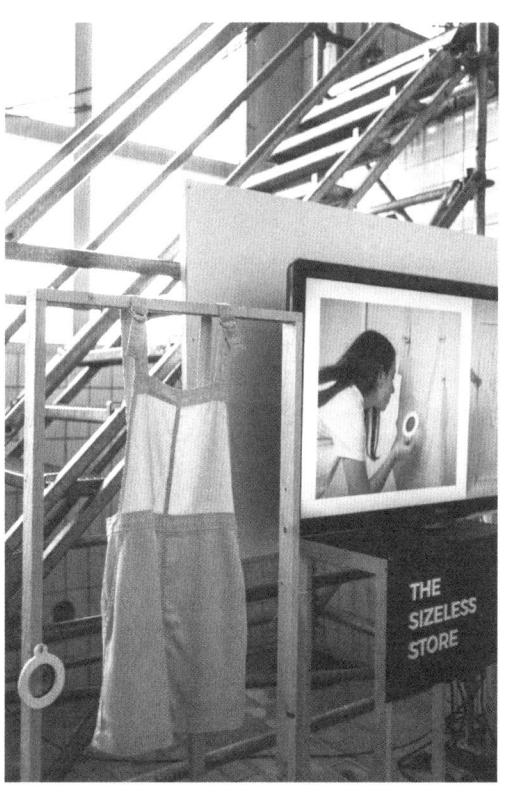

The Sizeless Store presented during the Fashion? Future design for the present, Dutch Design Week 2018

2018_44

The *Closed Loop Smart Athleisure Fashion* collection is a line of shirts for fashionable, sporty and forward-thinking women. The shirts contain technology that can measure your health, keeps track of the heartbeat and respiration. This technology is based on Holst Centre's advanced printed sensor technologies on flexible substrates for textile integration. The laminated sensors are designed for unobtrusive integration in both daily wear and conventional fashion production.

The lease and recycle system is a closed loop. So the user doesn't have to worry about what to do when the garment has no use anymore. In some pieces the recycled material development Econyl® is used. The technology will be delaminated from the shirt, and recycled. The closed loop system makes it a more sustainable way of producing, wearing and disposing of garments.

by-wire.net/clsaf/
youtu.be/6NHGZ1gooNM

Photos by Sanne Kortooms

**Closed Loop Smart
Athleisure Fashion**

**Holst Centre
StudioBonvie
by-wire.net**

Studio PMS is a fashion collective with a strong digital and innovative focus. Before they started doing digital design, they strove to form a countermovement against the hierarchy of their industry in a world that is screaming for sharing and interdisciplinary creation. The future perspective that PMS sees, is a progressive revolution in which overproduction and overconsumption can be reduced by creating tangible and perceptible fashion in a digital way. Their first project, *In Pursuit of Tactility*, was the start of their journey on digital gratification. This is becoming increasingly prominent in the fashion industry, as designers provide new methods to enjoy clothing in the digital realm. With their work they want to inform, stimulate and connect the current fashion industry and push their design-aesthetics further along the innovative and digital route. Shown project is made for Frame Magazine.

studiopms.nl/
studiopms.nl/img/trailer%20IPOT.mp4

Studio PMS

Puck Martens
Merle Kroezen
Suzanne Mulder

Digital designs for Frame magazine

2018_46

Hellen van Rees is a Dutch fashion and textile designer. She offers contemporary garments, based on traditional silhouettes and garment shapes, featuring unique textiles, hand-made fabrics, contrasting textures and innovative finishes. Typically, she combines craftsmanship with new technologies and unusual materials. She is an environmentally conscious designer who uses organic textiles and production leftovers. Aside from fashion, she also develops wearables in collaboration with the University of Twente and Saxion a.o. Here, the focus lies on creating textiles for haptic feedback used for coaching situations, such as a posture correction vest and a breathing trainer.

hellenvanrees.com
vimeo.com/206693657

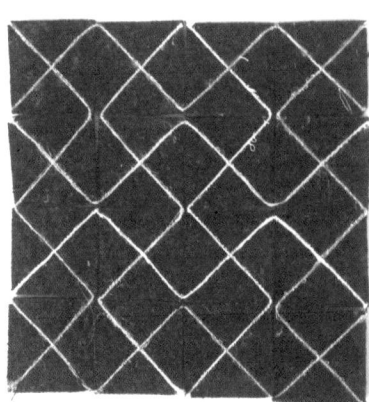

Detail of material

Textile Reflexes

Hellen van Rees

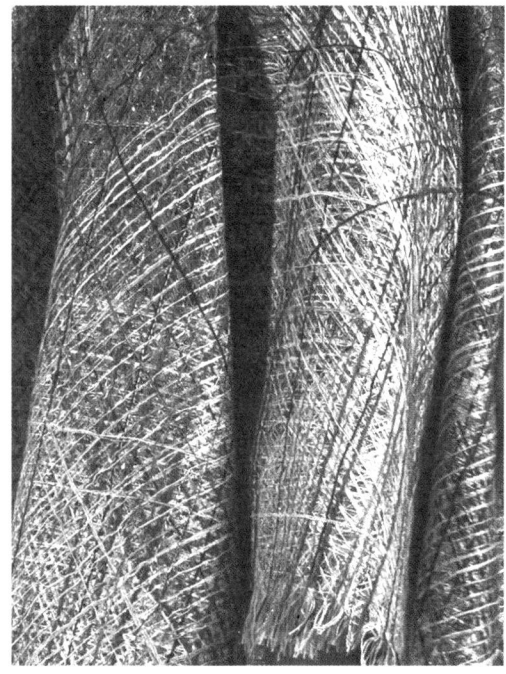

2018_47

Magic Lining proposes a garment that allows the wearer's the feel as if their body would be made of a different material. What happens in the transition moment, when the wearer shifts from his/her own body to the marble on, or the other way around? In the intersection of neuroscience research on mental body-representation (MBR), human-computer interaction (HCI) and real-life smart textile applications, the project ask questions about the meaning of clothing. In collaboration with: Ana Tajadura and Aleksander Väljamäe.

kristikuusk.com
vimeo.com/289294125

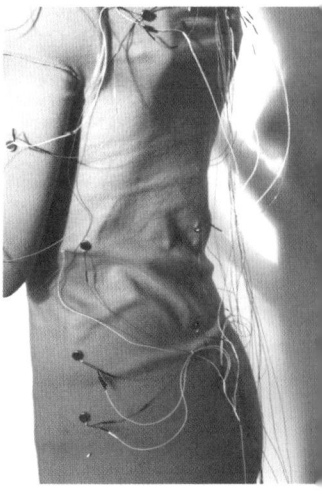

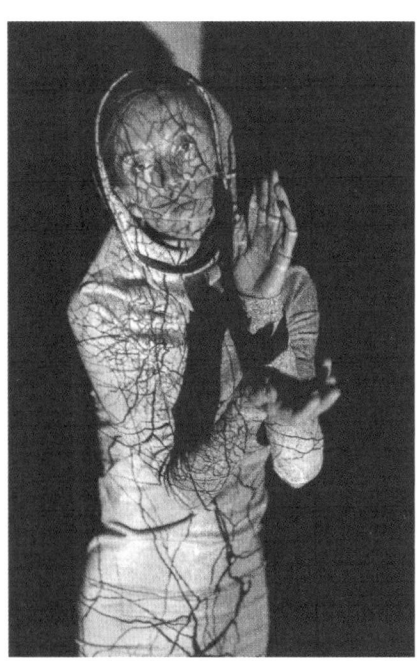

Magic Lining

Kristi Kuusk

Photo by Iris Kivisalu

2018_48

UNSEAM develops digital and on-demand garment production methods. Using simple digital techniques like lasercutting and laminating. Instead of sewing the few seams they are closed with a micromoulding machine developed by Bas. Using UNSEAM's own developed process based on material properties, programmed 3D shapes appear in the last phase of the production process. The reason why digital garment production is complex and expensive, is because textiles are difficult to handle with machines in 3D. This requires expensive robot technology. *UNSEAM* works in the flat surface using relatively simple and industrially available machines.

unseam.nl

UNSEAM

Karin Vlug
Bas Froon

Photos by Jeroen Dietz

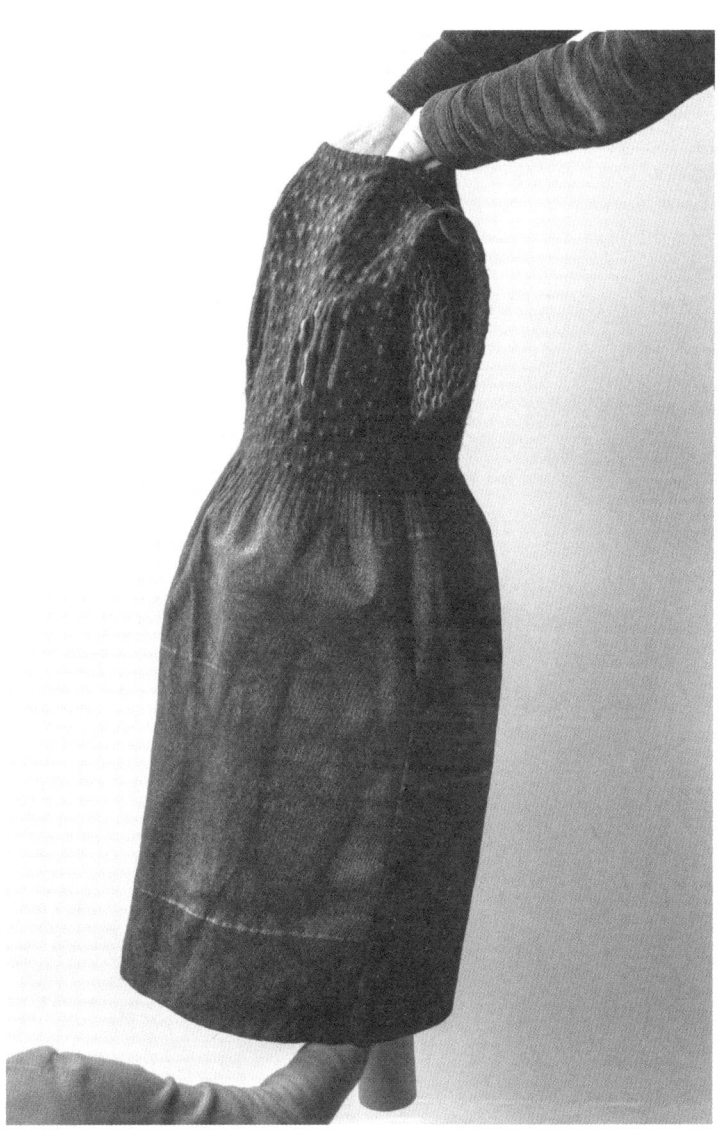

2018_49

Phem is a fashion brand concept for garments that use dynamic, surface-changing fabrics. It can also be understood as a design exploration for fabrics that can change imagery like that of a computer screen. It all began with us wondering, What would it be like to wear fabrics like this? and What would it be like to conceptualise and build a fashion brand that uses fabrics with these dynamic behaviours?

The fabric used by *Phem* works similarly to augmented reality in that it appears one way through a smartphone screen and another way in real-life. However, functionality was not the main goal. Making these garments and the fashion film that accompanies them was a way to explore what it means to wear and design garments with a hybrid digital-physical existence. Much like the way many of us live today—navigating between digital and physical experiences—we wanted to explore how this space could be used expressively for fashion. Can it be beautiful? Can it be feminine? Can it make us feel something? Can it make sense in everyday life scenarios, like for example, while reading the newspaper or drinking coffee?

phem.design
vimeo.com/312729991

Phem

Angella Mackey

Garments that use dynamic surface-changing fabrics.

conclusion

By Marina Toeters

2016_31, Agent Unicorn, Anouk Wipprecht

Over the last 50 years, innovation has been stagnant in the fashion industry in the production and process systems, and in the introduction of technology in the clothing products available in the high street. To change that, we need to collaborate, build new ecosystems and create systemic change. The approaches, tools and methods described within this book can bring about change in the fashion industry.

2009_04 Collaborative Textile: Fashion vs Interieur

In 2000, Phillips was at the forefront of collecting data on user preferences and the social impact of these innovations, and in overcoming many of technical and production-based barriers of public release. The end-2000's V2 initiative *e-textile workspace* was a regular meeting space for developers and designers within the field. The initiative supported the critical mindset and confidence of the participants, which included Piem Wirts, Melissa Coleman [2009_03], Ricardo o'Nascimento, Meg Grant [2012-15_13], and Anja Hertenberger [2009-14_05]. The Wearable Senses Lab at the University of Eindhoven also established itself as a hub for research and education. At present, all educational institutes in the Netherlands offer a Smart Textile or interactive and innovative fashion course, seminar or workshop, an example of which is by Beam Contre Choc [2018_41].

2009_03

2009_03, Mediavintage Charlie, Melissa Coleman

Global impact

The Dutch fashion tech community is relatively small, but Dutch pioneers are an important part of the global 'think-tank' in creating a better more sustainable future for the fashion industry. Meg Grant now works at Seismic (USA). She explains the difference between the USA and Europe:

> "I do feel like a lot of the really cutting edge stuff in e-textiles at the beginning of the 2000s was happening in Europe and a lot of it in the Netherlands. I think it's partly to do with the state support, but also because of who was interested in it at the time. Of course, there was stuff happening

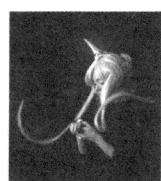

in Canada because of Joanna Berzowska, Kate Hartman, people like that, but I feel like the things that came out of the US were much more practical like Lilypad, Adafruit, Becky Stern, Lynne Brunning —Instructables!"

Dutch fashion tech designer Anouk Wipprecht [2016_31] educated in the University of the Arts Utrecht (HKU), the Netherlands, is now creating technological couture in Florida for clients like Intel, AutoDesk and Swarovski. Dutch designer Jesse Asjes [2009_04], [2013_15], [2015_13], also from the HKU, taught at the Rhode Island School of Design and now works in the USA as a knit designer for Nike. She suggests:

> "As I am carrying my Dutch Design DNA through my educational and professional background I can vouch for the Dutch Designers to be great collaborators. They highly value teamwork and it shows well by how they build on existing knowledge. There is no fear of sharing and building on new ideas together. To me they are vocal and confident in their competence and these are valuable assets to push boundaries. The Dutch Design DNA encompasses the explorative and curious manner of developing products and this resonates well with an innovative mindset. We are hungry for newness and finding elements to do things differently."

Jesse Asjes suggests that American designers have a great talent for marketing fashion technology and making designs commercially viable, but this can compromise in the simplification and effectivity of the designs. She states:

> "[Dutch designers] show a strong tendency towards critical making, therefore we sometimes oversee the strengths that are already accomplished. We could be stronger in celebrating what we have and build on storytelling around the existing work."

Similarly, Asian companies, whose biggest advantage is in having the largest hub of manufacturing on their doorstep, have mass production and sale of products as a primary goal. Thus, they have typically focused on the mass production of wearable fashion tech. Qi Wang [2015_27], assistant professor in the college of Design & Innovation in Tongji University Shanghai and previously PhD at TU/e, noticed this difference:

> "Dutch designers of wearable fashion are unique in the complexity of their research, as well as their capacity for deeper R&D. Most of all, they utilize empathy when developing prototypes and products; customized designs made to fit the individual bodies and wishes of its wearers."

Qi adds:

> "Chinese startups were positively promoting the manufacturing of various wearable systems and brought the affordable devices to the public with highly effective execution.

The European system is more conductive to this emphasis on R&D, with companies such as Philips being more aware and willing to invest time and money in long-term projects and market research. Still, Martijn ten Bhömer [2015_25], [2013-15_15], who did an PhD at TU/e and is now director of the master of design in Xi'an-Jiantong-Liverpool University, Suzhou China, and agrees with Qi, he says:

> "complex product innovation enabled by the rich research environment is great in the Netherlands, but we might want to learn more from the efficiency and speed of production processes in Asia."

The European WEAR Sustain program[1] funded almost 50 projects that involved a vista of new sustainable ways of designing and retailing fashion, as well as innovative production solutions [2018_45], [2018_47], [2018_49]. Ingrid Willems, the CEO & co-founder of DataScouts and consortium partner of the WEAR Sustain program explains her company's interest in such customized, empathically designed projects:

> "Throughout the WEAR Sustain project, it was clear that the Netherlands is a hotspot in wearable technology. The Dutch ecosystem of designers, academia, tech startups, mentors and experts involved in wearable technology is strongly interconnected with many links abroad."

2018_42, LABELEDBY.

2018_45 Closed Loop Smart Athleisure Fashion

1
WEAR Sustain (2017).
The WEAR Sustain
project has received
funding from the
European Union's
Horizon 2020 research
and innovation
programme under
grant agreement
No. 732098.

Towards the future

We are at a crossroads in the development of fashion technology. The active rate of development and innovation within the fashion tech pioneering community, the technology available to the producers and market, and the demands of consumers for change, provides an exciting vision for the future of the field.

So what's to come? UNSEAM [2018_49] developed an innovative local production process for garment making without sewing, the most labor intensive step in garment production. Perflex [2017_37] created fully customized garments by generative design via 3D-printing. This Fits Me [2015_29] uses 3D scanning technologies. LABELEDBY. [2018_42] combines textiles and 3D-printing and gained attention in production countries like Bangladesh. Angella Mackey explored the possibility of aesthetically and virtually animating garments using a smartphone app in her Phem project [2019_50]. Studio PMS [2018_46] went one step further by making the visual garments realistic and tactile without producing any physical object. Integrated technology that improves the way garments interact with the body like Zishi [2015_27] and Magic Lining [2018_48] will likely enter the market soon. The Closed Loop Smart Athleisure Fashion [2018_45] project showed that it is possible to create a fully customizable product with integrated technology that benefits the body, using sustainable production methods and innovative closed business loops through leasing systems.

The future is now, and fashion technology will play a significant part in that future by offering an alternative to the problems we face today. The pool of talent, knowledge and drive within the field is enormous, and it is my firm belief that it will not take long until fashion technology finally drifts into the mainstream. It doesn't matter who you are—designer, technologist, student or consumer— we all need you to be part of this changing paradigm. We hope that reading this book has been an excellent start or drive for your continuing contribution to our wonderful world of fashion tech!

2018_42

2018_45

index

epilogue

By Marina Toeters

This book wouldn't be here without the inspiration, design work, and willingness to share and participate of my colleague pioneers, many are represented in this book. Not all can fit within the limited space available, but nevertheless your work is highly appreciated. I am very proud and happy to see so many young people and former students represented in this book. This makes me confident that a bright future for fashion is inevitable.

Thanks to all the authors who contributed: Koen, Loe, Pauline, Lianne, Danielle, Rens, Ben, Jan, you have kindly agreed to share your valuable knowledge with the wider public through this publication and were a great source of inspiration during my years of working as a fashion tech designer in this field.

Alongside their contribution as authors, I would like to give extra acknowledgement to the following people: Stephan, thanks for sponsoring the base event *Fashion? Future Design for the Present* during Dutch Design Week 2018, planting the seed and supporting me to start working on this permanent reference and lasting overview document. Thanks Rens and Lucie for the support. Thanks Gail for your language contribution, this improved the quality to a great extent. Oscar, Anke and (yes again) Matthijs, thanks for being my critical reviewers, discussion partners, and title developers.

Daisy, thanks for always being my trustworthy backup during all my professional activities, thanks for being my critical counterpart, and in the development of this book, in which I was sometimes too emotionally evolved, you managed to make rational decisions.

Dearest Cindy, thanks for applying for your internship at the perfect moment in time! Your historical and theoretical expertise, your interest in fashion innovation, your outstanding writer skills and your pleasant presence meant it was a delight to have you involved in this process!

Thanks Eva and Freek for supporting this process in such an experienced and professional manner. Thanks to all of you who thought it was a good idea to develop this publication. It made its realisation a very enjoyable process.

epilogue

colophon

'Unfolding Fashion
Tech: Pioneers of
Bright Futures'
Onomatopee 167

Initiator, Producer,
Editor:
Marina Toeters,
by-wire.net

Production Assistance:
Daisy van Loenhout

Editor Assistance:
Cindy Tieleman

Text Editor:
Gail Kenning

Authors:
Daniëlle Bruggeman,
Jan Mahy, Rens Tap,
Ben Wubs, Loe Feijs,
Koen van Os, Gail
Kenning, Lianne
Toussaint, Stephan
Wensveen, Anke
Jongejan, Pauline van
Dongen and Oscar
Tomico

Graphic Design:
Inedition,
Eva van der Schans

Fonts:
Larish Neue
(by Radim Pesko)
and Graphik (by
Christian Schwartz)

Publisher:
Onomatopee,
onomatopee.net

© Marina Toeters,
by-wire.net

Print:
Wilco Art Books

Edition:
1600

ISBN:
978-94-93148-14-7

Powered by TU/e
Industrial Design

TU/e EINDHOVEN
UNIVERSITY OF
TECHNOLOGY

DEPARTMENT OF
INDUSTRIAL DESIGN

https://www.tue.nl/
en/our-university/
departments/
industrial-design/

Powered by NL
NextFashion & Textiles

NL NEXT FASHION
& TEXTILES

www.nlnextfashion-
andtextiles.nl
www.linkedin.com/
company/nlnext-
fashionandtextiles/

Detail *UNSEAM 2018_49* by Karin Vlug and Bas Froon